797,885 Books

are available to read at

Forgotten Books

www.ForgottenBooks.com

Forgotten Books' App
Available for mobile, tablet & eReader

ISBN 978-1-333-75899-8
PIBN 10544256

This book is a reproduction of an important historical work. Forgotten Books uses state-of-the-art technology to digitally reconstruct the work, preserving the original format whilst repairing imperfections present in the aged copy. In rare cases, an imperfection in the original, such as a blemish or missing page, may be replicated in our edition. We do, however, repair the vast majority of imperfections successfully; any imperfections that remain are intentionally left to preserve the state of such historical works.

Forgotten Books is a registered trademark of FB &c Ltd.
Copyright © 2015 FB &c Ltd.
FB &c Ltd, Dalton House, 60 Windsor Avenue, London, SW19 2RR.
Company number 08720141. Registered in England and Wales.

For support please visit www.forgottenbooks.com

1 MONTH OF
FREE
READING

at
www.ForgottenBooks.com

By purchasing this book you are eligible for one month membership to ForgottenBooks.com, giving you unlimited access to our entire collection of over 700,000 titles via our web site and mobile apps.

To claim your free month visit:

www.forgottenbooks.com/free544256

* Offer is valid for 45 days from date of purchase. Terms and conditions apply.

English
Français
Deutsche
Italiano
Español
Português

www.forgottenbooks.com

Mythology Photography **Fiction**
Fishing Christianity **Art** Cooking
Essays Buddhism Freemasonry
Medicine **Biology** Music **Ancient
Egypt** Evolution Carpentry Physics
Dance Geology **Mathematics** Fitness
Shakespeare **Folklore** Yoga Marketing
Confidence Immortality Biographies
Poetry **Psychology** Witchcraft
Electronics Chemistry History **Law**
Accounting **Philosophy** Anthropology
Alchemy Drama Quantum Mechanics
Atheism Sexual Health **Ancient History**
Entrepreneurship Languages Sport
Paleontology Needlework Islam
Metaphysics Investment Archaeology
Parenting Statistics Criminology
Motivational

THE ENGLISH COUNTRYSIDE SERIES
UNIFORM WITH THIS VOLUME

THE HEART OF ENGLAND
By IVOR BROWN

THE COUNTRYMAN'S ENGLAND
By DOROTHY HARTLEY

THE OLD TOWNS OF ENGLAND
By CLIVE ROUSE

ENGLISH VILLAGES AND HAMLETS
By HUMPHREY PAKINGTON

THE OLD INNS OF ENGLAND
By A. E. RICHARDSON

THE ENGLISH COUNTRY HOUSE
By RALPH DUTTON

THE CATHEDRALS OF ENGLAND
By HARRY BATSFORD and CHARLES FRY

THE PARISH CHURCHES OF ENGLAND
By J. CHARLES COX and C. BRADLEY FORD

THE ENGLISH ABBEY
By F. H. CROSSLEY

THE CASTLES OF ENGLAND
By HUGH BRAUN

THE SPIRIT OF IRELAND
By LYNN DOYLE

THE FACE OF SCOTLAND
By HARRY BATSFORD and CHARLES FRY

THE HEART OF SCOTLAND
By GEORGE BLAKE

Frontispiece

1 COAST VIEW NEAR POLPERRO, CORNWALL

Detail of a Water Colour by Philip Sheppard in the Victoria and Albert Museum

THE SEAS & SHORES
OF ENGLAND

By
EDMUND VALE

With a Foreword by
SIR ARTHUR QUILLER-COUCH ("Q")

Illustrated by Photographs

NEW YORK
CHARLES SCRIBNER'S SONS
LONDON: B. T. BATSFORD LTD.
1936

By the Same Author

SHIPSHAPE
SEE FOR YOURSELF
LOCAL COLOUR
THE WORLD OF WALES
ETC.

First Published, February 1936

MADE AND PRINTED IN GREAT BRITAIN
FOR THE PUBLISHERS, B. T. BATSFORD LTD., LONDON
TEXT BY UNWIN BROTHERS LIMITED, WOKING
PLATES BY THE DARIEN PRESS, EDINBURGH

FOREWORD

By SIR ARTHUR QUILLER-COUCH ("Q")

GUIDE-BOOKS to all parts of England, Scotland, and Wales multiply in these days; and the corner of England—the extreme West Country—to which I happen to belong—would seem to invite its fair proportion of the swarm, to say the least of it. Some of these do great credit to their compilers, mapping out routes for pedestrians and motorists, with distances, "objects of interest," architectural details of the churches, and just enough of geology or archaeology to be noted and talked about when the more curious visitor gets home. Others will take him along breezily, gossiping, advertising the town-dweller, be he motorist or hiker or invalid, that our seaside resorts provide for him a blend of natural beauty and ozone which at once braces the body and medicinally relaxes the mind with its cares; promises which, of course, are quite compatible. Others again advise concerning hotels, lodging-houses, entertainments, beaches, ways in which wise parents can get through the school holidays to best advantage.

Admirable as these books are, and efficient in their several ways, they have all—until I came to this one of Mr. Edmund Vale's—left me with a sense that they miss—or anyhow fail to communicate—almost half the joy which I, as a coast-dweller and descendant of coast-dwellers, have known and would share. They fix the reader's attention entirely upon the *land*, its villages, churches, country houses, parks, etc., with intimate details; they take him to the coast, encourage him to admire its rocky grandeur, but leave him to gaze over an expressionless sea; with a sunset, maybe, to help it, and some emotions. They never consider *the land from the sea*; never the beauty in the lessons of the coastline as read from that aspect; never even the exquisite mysteries of the inshore waters moving under the shore's foot. Yonder is the horizon; a straight line, as Euclid defined it, lying evenly between its extreme points, a plane superficies which "has only length and breadth," and—

>Is this the mighty Ocean?—is this *all*?

Now—save against those who plant their wireless sets and gramophones on the rocks beneath my garden—I have no grudge whatever against anyone who seeks the sea-shore on his lawful occasions, which are various enough. An eminent Professor, for instance, once drew me aside after a dinner party, to study a section of a maritime snail through his microscope. "The Polperro Snail," he announced, "a distinct species, and its habitat quite close to your own. But there—you have often observed it, no doubt?" Flattered, I had to confess that I had not. "But it breeds on the very face of the cliffs." And I had to explain that, familiar as I was with those cliffs, I habitually put my small boat about at a distance which forbade any close observance of snails. This for a few miles to eastward of my home. At about the same distance westward, a London magnate descended on the coast and, more or less like stout Cortez—exclaimed, or is reported as exclaiming, "The Almighty designed this bay for a Hotel!"—even as "my Father," in *Tristram Shandy*, once argued that Providence had shaped the human nose for the wearing of spectacles. With visitors as diverse in interest as these I can sympathise, as even with those who find health and pleasure in marking out courts on our beaches and playing lawn-tennis in the face of the Atlantic. Yet more do I enjoy the sight of children who, either naked or in bright vari-coloured bathing garments, play along the edge of the breakers "dance their ringlets, their ringlets to the whistling wind," or on yellow sands

>   with printless foot
> Do chase the ebbing Neptune, and do fly him
>   When he comes back....

But I would, if I could impart it, increase my fellows' delight by a vision of these same shores from offings; because while now in the late afternoon of life I can treasure memories of both—exploration of sylvan beauty which the motorist passes on his dull high road unguessing, though the dells start to plunge, right or left, within a few yards of the ridge and his speed—equal with these in beauty abide many hours and scenes of coasting: a sunset of flame upon St. Michael's Mount. My first opening of Falmouth harbour "under the opening eyelids of the morn," an approach to Plymouth under a heeling breeze as the town (now a City) suddenly illuminated, through the dusk, a face of welcome; or, on a shift of the helm well down past Ushant, a land breeze equally sudden, warm and laden with the scent of harvest-fields and orchards behind the amazing harbour of Brest. Add to such delights that of discovering just a little, yet more and more, of what the inshore fishermen know (but will grudgingly tell) of that unseen floor into which they so cunningly drop their nets; and I think the reader will surmise in Mr. Vale's pages a prospect of enjoyment, in a small way tested and proved by me, which he extends to limits far wider than mine.

ARTHUR QUILLER-COUCH

Dedication to
GRIFF AND DORA

★

ACKNOWLEDGMENT

THE publishers must acknowledge their obligation to the photographers whose work is represented in these pages, namely Aerofilms, Ltd., for figs. 19, 83; Central Press Photos, Ltd., for fig. 65; the late Mr. Brian C. Clayton for fig. 94; the *Daily Mirror* for fig. 92; Mr. J. Dixon-Scott for figs. 3, 5, 6, 9, 16, 21, 24, 25, 33, 36, 37, 38, 40, 48, 52, 54, 56, 57, 60, 61, 67, 69, 70, 71, 73, 74, 75, 82, 84, 88, 90, 91, 95, 110; Mr. Val Doone for figs. 2, 50, 51, 53, 77, 79, 80, 87; Mr. Herbert Felton, F.R.P.S., for figs. 18, 43, 86, 99, 101, 106; Fox Photos, Ltd., for figs. 7, 8, 72, 81, 103; F. Frith & Co. for fig. 34; Mr. F. A. Girling for fig. 100; Mr. Graystone Bird for fig. 63; Mr. Hemp for fig. 31; His Majesty's Office of Works for figs. 58, 97; Mr. Hutton for figs. 27, 28, 30; The Keystone View Co. for figs. 47, 68; London Midland and Scottish Railway Co. for figs. 13, 14, 117; Mr. Hugh Quigley for fig. 93; Mr. W. G. Sandy for fig. 10; Mr. H. Squibbs for figs. 39, 41, 42; Mr. Will F. Taylor, for figs. 12, 17, 20, 22, 23, 26, 29, 32, 35, 44, 45, 46, 49, 55, 59, 62, 64, 66, 76, 78, 85, 89, 96, 102, 104, 105, 107, 108, 109, 111, 112, 113, 114, 116; The Topical Press Agency, Ltd., for figs. 4, 15; and the *Western Morning News* for fig. 11. The Frontispiece is reproduced with permission of the authorities from a water-colour in the Victoria and Albert Museum.

CONTENTS

FOREWORD. *By* SIR ARTHUR QUILLER-COUCH ("Q")		v
ACKNOWLEDGMENT		vii
I.	ON BOTH SIDES OF THE LUBBER-LINE	
II.	THE IRISH SEA	10
III.	ST. GEORGE'S CHANNEL	38
IV.	THE SEVERN SEA	50
V.	THE ENGLISH CHANNEL	65
VI.	THE NORTH SEA	84
	INDEX	117

2 MORNING MISTS AT LYNMOUTH, looking to Countisbury Foreland, North Devon

3 A QUAY-SIDE AT POOLE HARBOUR, DORSET

THE SEAS AND SHORES OF ENGLAND

CHAPTER I

ON BOTH SIDES OF THE LUBBER-LINE

WHERE the sea is concerned, there is a fundamental difference between the points of view of the landsman and the sailor. To the former, the sea is a happy adjunct to the land, causing it to form cliff scenery and bathing-beaches. To the sailor, the land is an unhappy intrusion from the sea-floor, marring the beauty and pathfulness of his beloved element with ugly threats of danger. But there are two kinds of seas—high seas and narrow seas—and it is with the latter that we have to do in this book. A narrow sea has an individuality of its own. The landsman notes it by the character of its coastline, by the trade of its ports, and the racial attributes of its littoral inhabitants. To a sailor, any sea you might name (supposing he knew it) would conjure up an embodiment of tide-rips, overfalls, soundings, prevalent wind-drift, fog-banks, landfalls, and "departures." He would have a certain definite idea about the colour and the roll of the water, and certain kinds of sky would probably occur to him as inseparable accompaniments to the picture, skies that you only got in the one sea mentioned and in no other throughout all the broad hydrosphere.

For our purposes I would like, if I could, to strike a happy medium between the partiality of the landsman and the partiality of the sailor, and strike it just so nicely as to get the essential spirit of each sea under discussion without becoming too deeply involved either in geography or its even more laborious antithesis, oceanography. For I believe that each of our five British seas (to go no farther abroad) has a distinct personality of its own, a personality more subtle and influential than we are generally aware of. Our historians have taken no pains to enlighten us on this point. And where will you find the schoolroom where Admiralty charts are available for study as well as maps? Yet all the familiar things we hear of continually about Britain, from its secret pride in Empire to its public pride in insular idiosyncrasy, can be traced quite as much to the nature of our five seas as to the happy accident of geology which has made the homeland.

The pageant and the cinema have given people a natural, if unacademic, education in the leading types of costume of various periods which is of the utmost value to the sightseer when he visits a mediaeval ruin. For he can automatically people doorways, stairways, battlements, aisles, choirs and cloisters with lifelike figures, and enter into the spirit of the old building whose secrets, lessons, parables and predictions for the future he would challenge. But it is not by any means so easy for anyone unversed in coastwise lore to reincarnate the spirit of a small ancient seaport or fishing village that is now in the process of turning into a seaside resort of uncertain grade. Yet we have only to go back twenty or thirty years to find these places in the full bloom of the romance which time and history had made them heirs to.

I will try to fill out some of the gaps which modern conditions in fishing and seafaring have imposed. Let us go back into the 'nineties, when motors were scarcely known on the roads and completely absent from the sea. In those days a small harbour had certain fine flavours of smell, sound and sight which made it absolutely distinctive from the inland town or village. You could be brought into it blindfold and yet make no mistake. The flavouring which greeted your nostrils was that of tar, new cordage and new shavings, drying canvas, drying nets and marine gear of all sorts. These things did not rejoice the heart because they smelt of themselves but because they were as pepper and salt to the sea air and brought out its peculiar allurements. There were incessant sounds of the beat of oars, the creaking of block-sheaves, the clack of windlasses and capstans, of the mallet on the caulking-iron, of the adze and other particular tools, all of which seemed to make a peculiar and delicious blend with the sounds natural to the sea itself, waves, mist-echo, the mewing of gulls. As to the sights, they may be mostly imagined when one remembers the tradesmen that were then indigenous to every port, the ship chandler, the sail-maker, the rigger, the shipwright, in addition to the sailing-ships, the sailors, the watermen and the fishermen.

In those days the tiniest places not only built boats but even ships. Take Llangranog, in Cardiganshire, which had no "facilities" in the matter of dock or slipway—there is not even a quay-wall there. It can only be called a port because

4 FROM THE DECK OF A NORTH SEA TRAWLER

5 MENDING THE NETS AT HASTINGS

6 OLD-TIMERS AT PORTH LEVEN, CORNWALL

it has a village with a church and a public-house down on the very beach itself, beside a stream which issues through a cleft in the cliffs and meets the sea on a bottom of fine, hard sand. Nevertheless it was a port (though not in the Customs sense), and about the time whereof I write they built a full-sized brig here, and launched her, making skilful use of a spring tide and the scouring properties of their little stream. Moreover, several of the small population found their work in her, and others lived happily ever after on the shares in her earnings, for it was customary to build and own a ship jointly—in 64 shares. They had, of course, other visiting ships which brought them their provisions, their coal and their lime, and men, boys and girls engaged in an energetic fishing industry which had gone on from time out of mind. Although they lay practically open to the Atlantic, they had never thought of installing docking or wharfage facilities, far less had they ever thought of relying for support on summer visitors.

The summer visitor, with all the paraphernalia to accommodate and amuse him, is now far more in evidence at coastal places than the seaman and things maritime, so much so that at a casual glance it would seem as if all that fine building up of courage, resource, industry and thrift which the peril of the sea had inspired and consolidated was dissipated and become less than a ruin. But nothing sticks like tradition. If you go to the seaside to enjoy all those strange and exotic pleasures which publicity brochures advertise, you may or may not find them in a town which has a long seafaring tradition. But if you go to enjoy the sea and the things of the sea you will only find these joys in a place which has served its time to sea hardship through centuries. There you will surely find them, in spite of a heavy camouflage of esplanade, winter gardens and all the raffish junk of tripperdom, even in spite of the distracting bungaloid fungus. But you must have a trained sea-consciousness to penetrate the modern disguise.

But I was talking of the 'nineties, when sea-consciousness was not an acquired taste but something which forced itself on the notice of all visitors to the seaside. The horizon was hardly ever a blank in those days, except in parts of the East Coast where there is no deep water near the land. Sail of all kind, from the full-rigged ship to the fishing-smack, went in continuous solemn procession from right to left and from left to right, or appeared and disappeared over the arc of the world. In storms, wreckage was flung up everywhere along the beach, and the lifeboat gun was a common sound. The whole population of a place would dash to see the launch and behold a boatload of men battling with oars against the bursting seas. Hardly a day passed but you would see the naval coastguard with his trim beard and smart sailorly appearance ranging along his well-trodden beat with his telescope tucked under his arm. And there was no weather report then except from the mouths of certain oracles on the water-front. And the stories that were rife then from the old hands of the port, how well they beat all the outpouring of modern fiction! And you might say then of every ship chandler, sail-maker, ship's carpenter, bo'sun, mate and captain that he was a "character." But the streets at night were full of drunken men, for the average sailor's life was no better than that of the average cab-horse.

And now the sailing-ship has gone, together with the picturesque premises which owed their existence to her, the chandler's shop and "marine store," the sail-maker's loft, the local slipway. The shipping is all in big tonnage which cannot enter the small ports and is concentrated in the big ones; so is the fishing. There are few "characters" of the old school left, few weather-prophets, and few drunken sailors. But although it has lost its best exponents, the mystery of the sea remains. And in its new detachment, shorn of so many distractions and fixed expertism (for the old sailor's contempt of shore people who held views about the sea was very dashing) the landsman is left a greater latitude both for theorising on and for venerating it.

But, as I have said, our task is an even more particular and peculiar one than that which occupied the sailormen. It is to deal with the several narrow seas which are called by the names of the Irish Sea, St. George's Channel, the Severn Sea, the English Channel and the North Sea, and to deal with the home coastlines of these.

Now, although there have been no true sea historians in the sense that there have been land historians, there has, for the last half-century, been an organised band of sea topographers whose works are still almost unknown and utterly neglected by the landsman. The practical findings of this band are reissued from time to time in a series of publications known by the general name of Sailing Directions. Individually, each book is called a Pilot—the North Sea Pilot, the Channel Pilot and so on. It is the earlier editions of these books which are of special interest as they were addressed exclusively to the mariner of wind-driven ships and are largely expressed in that most delightful of all professional jargons, nautical language. In addition, their authors spared no pains to find out all they could discover about the coastline, not only its configuration and the nature of its harbours but even matters of history relating to old buildings which stood out as landmarks. The old Pilots contain a quantity of information and coastwise gossip that has been preserved in no other form. To us landsmen it has a particular value, in that it is all given from the nautical point of view.

The sailor has not only his own topography but his own set of place-names which may or may not be acknowledged in those locations ashore. Sometimes they are quite unknown, sometimes they have succeeded in ousting the landsman's own nomenclature and have received the final test of recognition by appearing in print on the ordnance map. As examples of this I would cite North and South Forelands, Cape Cornwall and Holyhead. Of these, only the last has a remembered local original, for Holyhead is properly Pen Nhaer Gybi (the Eminence of St. Cubi). But most of the Welsh headlands were forcibly Anglicised from the sea, as are the foreign capes, such as Ushant (properly Oessent). A fair example of this is Point Lynas in Anglesey where the Liverpool pilot is picked up. It is named after the local saint Elian. Elian Latinised becomes Elianus. The transference from the Celtic to the classical form seems to have been made when a pretentious mansion was built on the headland in the early nineteenth century. Some time afterwards the Mersey Docks and Harbour Board acquired this mansion and converted it into a lighthouse. Probably before the date of this conversion the point had one name for the landsman and another for the sailor. Now, however, the name of the luminous mansion of Elianus was bound to succeed. But the sailor would learn it not from print but from the local man who, being a Welshman, pronounced it Elianis. Hence it became fixed as Point Lynas.

Other odd things have occurred. The old church at Reculver in Kent had twin towers at its west front. Hence, to the sailor, it has ever been known as The Reculvers. On the Lizard, in Cornwall, there used to be two towers, those of an old and a new lighthouse. Sailors therefore named that headland The Lizards. And though the old tower has long ago disappeared no true old salt will ever be persuaded to name the headland in the singular.

Of greater interest, however, is the submarine topography, which is invisible to the landsman but always as keenly apprehended by the sailor as if he saw it. It consists of channels, sandbanks, and submerged rocks. This topography is one of the principal elements which gives a narrow sea its entity and personality. The subject has never, so far as I know, been "done" in literature, any more than sea history has been "done." Yet if the names of this submerged region were collected, sifted, analysed, and compared, they might yield several missing links to our established history and romance, for many of them go back clearly to Saxon and Viking days, and it is likely that Roman and even ancient British origins might be found. The names commemorate wrecked ships, captains, merchants, saints, events, resemblances. And one may suspect that the latter class would yield links in mytbology. As a case of how a man may know his sea I would cite a captain in the Heysham–Belfast service who can tell, when he is lying in his bunk, by the sound which his ship makes as her hull is thrust through the water, exactly where she is at any moment of the voyage. To this man the Irish Sea is not a mere blank of salt water surrounding the Isle of Man and her adjacent coasts but a highly unique complex of shoals, spits, banks, patches, grounds, channels, swatches, deeps, lakes, gulleys and guts, all varying momently in their aspect of friendliness or menace according to the state of the tide.

But now, let us make some close-up observations on the coast itself. As a theme for holiday admiration, for aesthetic contemplation, for poetry or the colder philosophies, it may be divided under three heads—

    The Offing;
    The Shore;
    The Entrance.

The *offing* is that part of the sea which lies immediately off the land—say as far as the horizon line. An offing may have three features which are of joint interest to the landsman and the seaman, namely, when it contains a roadstead, a fishing-ground, or if it is a highway for shipping. A roadstead is a place where ships may anchor to await a favourable tide for entering harbour or favourable weather to proceed on

7 A VETERAN OF THE IRISH SEA: the Barque "Killoran"

8 COALING SHIP: Up-to-date Methods in a North Country Port

their voyage. It must have a sandy bottom in which anchors will take firm hold. With the passing of sail, most roadsteads of shelter have now little interest for the beholder except what is historical and sentimental. In that most famous one of all, the Downs, it is a rare sight to see a wind-jammer hove-to waiting for a favourable breeze to proceed up or down Channel. Yet as lately as the beginning of the century they lay there by the hundred assiduously attended by the Deal luggermen who touted for commissions to pilot, to bring off provisions, to take hands ashore.

Whether you believe in ghosts or not, the mind's eye will present you with something very like them if you charge it with the right material. And an offing seen from a cliff-top on a moonlight night is a background on which the figures of the past may be conjured with unusual facility, as nothing in it has ever been altered or modernised. With the same sounds in your ears and the same tang of brine and wrack in your nostrils that stimulated the senses of the watchers in the past, you may look out on the same kaleidoscopic textures of gleam and sheen and shadow on which the lone black figure of a ship appeared and prepared to make a signal. The smuggling lugger! We may safely raise this friendly spectre in most of our offings.

For the fisherman, the offing has an even more individualised topography than it has for the sailor. It is full of "marks." These are cross-bearings—to give them a more technical name. For instance, if you get the church tower in line with the second fir-tree above Mr. Hodge's farm, at the same place where the bush on top of the quarry appears just over the lime-kiln, you are hove-to over one particular spot on the sea-floor which you may always find again by remembering the said marks. Local fishermen may be ignorant and unlettered men but they have a memory for their special marks which a scholar could not compete with. These are the true secrets of their trade, and they guard them jealously, as well from each other as from the outsider. The line-fisherman has marks for feeding-grounds (for some fish are as conservative as restaurant-diners). The small sailing trawler which fishes within the three-mile limit has marks for "fasts," which are old anchors from ships lost in times out of mind, and wrecks. A man who does not know the marks of the local fasts will soon rip or lose his trawl gear.

By *shore* I mean the region of that invisible line mentioned on maps and charts as H.W.O.T.—High-Water-mark Ordinary Tides—and is made to include the cliff verge as well as the beach of sand, gravel and shingle, and the mud-flat. It is a mistake to take the shore for granted, to think of the beach only in terms of sun-bathing, and the cliffs only in terms of spectacular scenery. If you have an eye for long perspectives in time you will see the shore as a ceaselessly active factory in which the landscapes of the future are being made out of the landscapes of the present and the past. The cliff is the child of an ancient beach and will be the mother of new beaches within the powers of our prediction. In turn, the new beaches will be mothers of other cliffs in ages outside the compass of human anticipation.

Our cliff scenery is probably the most varied in the world, as the Isles of Britain seem to have been singled out by geology as a concentration-point of representative exhibits from nearly every period in the world's known rock history. And every rock-formation has its own technique in cliff-making, both as to colour and form, ledge-making, cave-making, and beach-making.

Sand is found most abundantly in the neighbourhood of sandstone rocks which occur at the three geological epochs called Triassic, Carboniferous and Devonian. It may reach the sea down river channels which course through these rocks, or from the cliff-face. Then its dispersal depends on the run of the tidal current along the coast and on the tendency of the coastline to form bays. A bay interrupts the drag of the current and causes the charged water to deposit its burden of sand. In many places this action is assisted by the placing of wooden groins at right angles to the edge of the water. Much sand, however, is made of pulverised shells, but this tends to blow inland and form ranges of sand-dunes.

By an *entrance*, my third heading, I mean any point in a coastline where human contacts can be made between the sea and the land. Without artificial aid this can only be done where rivers are deep enough to carry ships up into the land or where they afford estuaries at their mouths, or where the rock rifts to let in a fjord or a creek. By artificial means, however, in the shape of moles and harbours, bays and even open spaces can be adapted to form entrances, but many of them are unsafe in certain winds.

The study of an inland town is based on its market, its manufactures, and the nature of the surrounding country.

9 SAILING TRAWLERS AT LOWESTOFT, SUFFOLK

10, 11 THE " H EVANS" ON A LEE SHORE: the day aft

But the study of a seaport is based on its trade and the nature of its entrance. An inland town can subsist on manufactory alone or on a market alone. But a seaport town cannot subsist on harbour and docking facilities if there is no trade, or on trade potentialities if the sea entrance is not made efficient. An inland town is bound by county traditions and prejudices, but a seaport is linked with the whole world of land and also with that mysterious other world of the high seas which only the initiate can discuss among themselves. So the population of the seaport is inherently different from that of any other place.

Although we boast much of our heritage as a maritime nation, we were exceedingly backward in the matter of making harbours compared with foreign countries. In the mid-eighteenth century there was not one good artificial harbour in the whole realm. At the beginning of the nineteenth century there were only two indifferent docks in London, transhipment being almost entirely conducted by lightermen from lines of ships anchored in the river. Yet the countryside depended far more in those days on receiving all kinds of goods by sea transport, fetching the things from the nearest point on the coast where they could be dumped so that practically every little creek and river-mouth had its port and flourished until the coming of the railways, when the coastwise trade slowly but surely declined. The motor lorry dealt the final blow.

Just as the general historian scamped his duties by the sea, so the local historian seldom does justice to the small ports, though their stories would generally be more interesting (if told by the right person) than those of the county town and cathedral city. It is left to the sightseer who has a keen eye, a quick ear, and a sea sense to find out the true lore of the coasts of the seas of Britain—for the following pages are only a very general framework.

CHAPTER II

THE IRISH SEA

THE Irish Sea is the most truly British of all seas. For if by British you mean something pertaining to the Isles of Britain, the Empire of Britain and so on, the modern meaning of the word, in fact, which takes no account of Celtic origins, you could not find a more representative concentration of Britannic elements anywhere. The shores of the Irish Sea are parcelled out as if by some impartial arbiter into holdings of almost equal size for all parties—the Welsh, the English, the Scotch, the Ulstermen, the Free Staters, the Manxmen. Or, if by British you mean it in the stricter sense of *Brythonic* the statement holds good, for it is the sea of the ancient kingdom of Strathclyde as well as the sea of North Wales. In fact, the richness of the Celtic twilight in the Irish Sea is only surpassed by the richness of its material sunsets. Turner was a regular visitor to the Dee estuary for the sake of the sunsets in Liverpool Bay, and there is an afterglow in Morecambe Bay which fires the whole assemblage of the Cumbrian mountains and throws their image, beacon-red—but showing every outline, on the wide shallow waters and desolate miles of sad wet sand-flats.

The Irish Sea is the most enclosed and, at the same time, the most picturesque of all our seas. The six countries which surround it and the one which lies mid-most in it have nearly all contrived to throw up the grandest and wildest of their mountains along its shores. Here muster the Snowdon Range, the Clwyd Hills, the Lancashire Fells, the mountains of the Lake District, the Galloway mountains (Cairnsmore of Fleet is the wildest and most imposing mountain in the Scottish Lowlands), the Manx Mountains, the Mourne Mountains, and the Wicklow Mountains.

I feel certain that if this sea had been anywhere else but in Britain we should have "discovered" it long ago. We should be singing its praises, writing its romances, touring it, cruising it, publicising it. As it is, we hardly think of it except in connection with the four famous crossings to Ireland, all of which, with one exception, are made in the night-time. Yet if you come to sail this water in a small boat

12 HOLYHEAD, ANGLESEY: the South Stack Lighthouse

13 THE IRISH MAIL-BOAT OFF KINGSTOWN HARBOUR

14 ST. PATRICK'S CHURCH, HEYSHAM, LANCASHIRE:
a Saxon Ruin overlooking Morecambe Bay

you wonder why all the cruising yachtsmen have not concentrated on it. But yachtsmen like rock-climbers are very much "birds of a feather."

If you want to find the unknown countries of the Irish Sea it is better to adopt the method of Columbus rather than that of Messrs. Borrow and Morton, and go by boat. These unknown patches lie to the west and north. For County Down in Ulster is still very nearly as it was when I remember it in 1905. You need some sailorly hardihood to enter Strangford Lough (which retains its apt Viking name of Strong Fiord). The narrow entrance is about nine miles long, and the tide races in and out with a top speed of six knots. So fierce are the eddies that you would think shoals of porpoises were splashing round your barque when you arrive through the narrows into the first wide portion where the two truly old-world towns of Strangford and Portaferry face each other, each with its ancient quay and castle.

After that, you pass into an immense salt lake surrounded by a rolling landscape of pasture and cultivation and woodland which carries its trees and hedges down to the water's verge. So large is this lake that it gives you a water horizon with the far land only dimly looming over it. There are many islands with tolerably large farmsteads on them. On one is Nendrum, the only Celtic monastery that has ever been fully and scientifically searched by excavation.

A little farther south is Ardglas, a port well known to the Scottish herring fleet but not to many tourists. It has the ruins of seven individual castles, and its immense harbour works, which are deserted though not in ruins, belong to those strange days which now seem as far off as mediaeval times, when optimistic British Governments built harbours in Ireland rather than in England, Wales, or Scotland.

Dundrum (father south again) is another place which can only be discovered from the sea. The town with its splendid shell-keep castle lies in a spacious lagoon which has an extremely beautiful and romantic entrance through a vast range of sand-hills, an entrance that but for the presence of an old black buoy (which is not where either the chart or the sailing directions tell you) would be as completely invisible from the offing as the entrance to fairyland. Never have I found such fine cockles as in the sands of Dundrum. But perhaps these sands have some peculiar virtue, for it is said that the famous Hugh O'Neil, of Elizabethan fame, when he had been banqueting well but unwisely up in his castle, would have himself buried up to the neck in them (if the tide were all right). Spending the night like this he found to be an excellent antidote for a thick head.

The perfect round flowing shapes of the granite mountains of Mourne, too, must be seen from the offing of Newcastle Bay if you would have this vision at its best. I have never seen a finer group of granite hills anywhere in the course of my travels. When the Irish air takes on that vivid clearness in a false lull before the sou'wester, those hills bloom into a melting purple as soft and sweet as their contours—you feel you can almost taste them.

Then comes another fiord. This one the Vikings when naming it Carlingf(i)ord must have felt truly at home about. For it is spectacularly mountainous, having the smooth Mourne to the north and a precipitous craggy range to the south. Unlike Strangford, it has a wide straightforward entrance. Warrenpoint is on the Ulster side, and Carlingford on the Free State shore. At its closure stands Newry. Greenore is on a spit of sand at the southern point where the wide, shallow Bay of Dundalk begins.

The pastures of Meath and Louth are probably the most luxuriant in all Ireland, and truly lush are the banks of the River Boyne, which is still navigable for steamers up to a thousand tons (provided they have been built so as to clear the bar) as far as the ancient double town of Drogheda where Cromwell got himself that bad name that is still in circulation as a special curse.

With sandy intervals at Skerries and Balbriggan (the particular port for Lambay Island) the coast goes uneventfully down to Dublin Bay where the Wicklow mountains begin their ascent, and I will maintain with anyone that there are few finer landfalls anywhere than Kingstown Harbour as seen from the deck of the Holyhead mail packet at five o'clock every summer morning, with the mountains, headlands, and islands of Dublin Bay touched by the rosy-fingered dawn and set to music by the surge of the bow-wave and the mewing of the gulls.

Before speaking of the north shore of the Irish Sea (where I promised discoveries as unsophisticated as those in County Down) let us turn to the Isle of Man. The Isle of Man is a compact essence of the spirit and the tradition of the Irish Sea. In its ancient monuments it supplies forms which are unique and are yet links between the more familiar and more studied examples of England, Wales, Ireland and Scotland. Such is the stone circle of Craigneish, which far more people of antiquarian tastes would visit if it were not for the tripper bugbear. Yet in the very height of the holiday season few trippers find their way to Craigneish although the place has such curiosities as the last fluent-speaking Manxman and the best examples of Manx thatching in the island. Also there are archaeological pointers to show that the village has remained in the same place since the Stone Age (say between three and four thousand years). Practically every male who lives there is engaged in the great and permanent industry of the whole island—going to sea. The Irish Sea is the universal provider of employment all round the coast. For if you are not a sailor or a fisherman you can be a fisherman-farmer, and this occupation yields the most romantic type of smallholder conceivable, and one who usually manages to make ends meet.

There is an unusual dash of the nautical about the ancient monuments too. The only ship-burial known in Britain is here. In the museum at Douglas you may see not only the remains of a sea king but the remains of the very ship he cruised in also—such as have survived corrosion. And one of the high crosses has a contemporary picture on it of a Viking ship with sail hoisted.

The Isle of Man is both mountainous and dead level, the break happening suddenly and absolutely. In fact the whole physical geography of the island is entertaining and dramatic. It proceeds from south to north (magnetic) thus. A lighthouse rising out of the sea itself from a rock pedestal that is scarcely discerned called the Chicken. This Chicken gives its name to a formidable and far-flung tide-race which was of yore one of the main terrors of the Irish Sea before the days of steam and wireless. The Chicken is followed by the Calf. This is an island, and the zoomorphic imagery is not helped by its full name, which is the Calf of Man. It is a picturesque heather-crowned islet that supports one farmstead. Between this and the main island there is a passage called the Calf Sound, where the tide-rip displays all those features of terror and allurement described by Stevenson so well in *The Merry Men*. There is an unwritten bookful of shipwreck stories about it. Then you have the mountains from South Barule in the south to North Barule in the north, which is neighbour to Snae Fell, the highest point in Man (2,034 ft.). This march of mountains northward, having achieved the two thousand-foot contour, ceases as suddenly as if someone had cried "Halt!" The land then proceeds northward to the Point of Ayre as flat as a slightly crumpled pancake, which is known as the Plain of Andreas.

I must confess to a weakness for the Plain of Andreas. Those crumples in the pancake I have mentioned which were produced by flood waters at the end of the Ice Age are a topographical feature of great interest and fascination. Needless to say the hill-folk of Man despise these lowlanders, and the inhabitants of Silby, Jurby, Andreas and Bride have not a good word to say about any of the Manxmen to the south who have the misfortune to live in elevated country.

Although the Isle of Man is both ethnologically and nautically the hub of the British Isles it has only been able to assert its position at two stages of our history. The first was when the Vikings made it a base from which they founded the trading ports in Ireland which included Dublin, Wicklow, Wexford and Waterford while they launched raids on the coasts of Scotland, England and Wales, forming colonies there, though none which were destined to last so well as the Irish foundations. But there were Scandinavian kings of Man even after the collapse of the Norse influence in the Irish trade ports owing to the Norman invasion of Ireland. And it is an odd fact that the daughter of the reigning Sea King of Man at the time married De Jocelin, the conqueror of Strangford, an indication that blood is thicker than water, for the Normans were themselves Norsemen with an adopted veneer of Latin culture.

The second period of Manx supremacy in the Irish Sea may be dated as between 1670 and 1765. The first date covers the arrival of a company of merchant adventurers who emigrated from Liverpool to Douglas for purposes of organising the contraband trade from that centre. The Isle of Man was then a small "private" kingdom held by charter, and in the hands of the Stanley family, Earls of Derby. It was practically duty-free, though I suppose the lord of the manor saw to it that he benefited as well as his subjects. It was ideally situated, not only as a distributing centre but also for the reception of goods. As it was perfectly legal to land goods here, such splendid argosies as the East and West Indiamen of Bristol and Liverpool found it well worth their

HARVEST TIME IN THE ISLE OF MAN

16 CASTLETOWN, ISLE OF MAN, AND CASTLE RUSHEN

while to call with tobacco and silk on their homeward voyages. As duties were increased the business became more and more brisk. There were even offices set up in Liverpool and Glasgow in connection with the Isle of Man "merchants," for one end of the business always remained legal. At the other end of it were some of the cleverest and most daring seamen who ever sailed the waters of the Irish Sea.

These Irish Sea smugglers, whose history has never yet been written, were mostly Manxmen, though by no means all. In large but very handy luggers they ran their cargoes to a hundred secret trysting-places round all the coasts of the Irish Sea and to other places through the North and St. George's Channels. They had all the shore population, including parsons and even magistrates, on their side, and it was long before the Government preventive service was able to compete with them.

Of all their routes, the favourite and the hardest worked was the trip to the north shores of the Solway Firth. It was the most efficiently served on the shore end, and, with luck, could be made out and back in the course of a single night. And there are probably more caves, hiding-holes, trysting-points and farm-houses friendly to the free-traders, together with their appropriate stories, that can be still traced than anywhere else. Old ballads, too, still run, as:

> Aul Johnnie was a pawky loon;
> A pawky loon I trow was he;
> At smuggling on the Solway coast,
> Bune ithers aye he bore the gree.

The coast is indeed all that a smuggling romancer could wish for. It begins with the Mull of Galloway standing like a clenched fist in the sea with cliffs rising sheer to 250 feet. Then come the wide, shoal waters of Luce Bay and, round Burrows Head, a second immense inlet, Wigtown Bay. These waters, with their great forlorn expanses of sand and runnel and their swift moaning tides fit the tragic story of the women who were pegged out at low water to perish slowly, at the time of the Covenant martyrdoms. The rocky shores of Kircudbright and the romantic estuaries of the Dee and the Nith with their wild moor and rock background always dominated by the mountain, Cairnsmore of Fleet, offer a subject that is tempting but, alas! too complex to deal with in our hasty survey. It is a lovely coast, still instinct with that magical force of environment that lifted Sir Walter Scott to his best work. And the memory of Dirk Hattrick and his comrades is, after all, a trivial and sordid one beside that of the lady of Sweetheart Abbey and the lord of Carlaverock Castle and the knightly trysts at the Lochmabyn Stone, each a clear field for fiction writers yet to come.

The headwaters of the Solway have a flat shore on the English side composed of great salt marshes on which hosts of store cattle pasture. But, like the Plain of Andreas, this lowland has an abrupt end a few miles inland, when it soars steeply to one of the chief eminences of the Lake District, Skiddaw, which achieves 3,054 feet. Macaulay's poem rings heroically enough, and one hopes his facts were in order when he says that—

The red glare on Skiddaw warned the burghers of Carlisle.

But, talking of these burghers, it is worth noting a brief effort made by them to place themselves upon the sea-coast by owning an out-port. In the early nineteenth century when canals were all the fashion, they made a canal to the Solway and opened a new haven called Port Carlisle. One pictures the usual optimisms, the cutting of the first sod with a highly decorative spade, the speeches on opening day. But then the canal did not bring the expected results in trade. Very shortly after, canals went out of fashion and railways came in. The course of the burghers of Carlisle was once more made plain. The canal was filled in and a line of railway was placed on top of it. Even then things did not prosper with the outport, and shortly afterwards its fate was finally sealed by the Glasgow and South-Western Railway which actually built a viaduct between the port and the sea through which nothing but a rowing-boat could proceed.

But the South Solway has an industrial area though it lies in a direction remote from Carlisle—towards the sea, that is. It is represented by the four ports, Silloth, Maryport, Workington, and Whitehaven. Silloth, the highest up the firth, is not greatly concerned with industrialism. In fact it gets the general trade which was expected at Port Carlisle. And a very typical Irish Sea trade it is, supporting a regular steam service to Douglas, Dublin and Liverpool. Maryport is largely concerned with fishing. Workington and Whitehaven are exclusively coal and iron. They are the chief outlet of the Cumbrian coalfield.

The corner-stone by which the Solway Firth joins the open sea is well marked by the three hundred-foot cliff of St. Bees Head. It is made of red sandstone of the Triassic Period, and is one of the few effective sea cliffs which this geological epoch has succeeded in producing. But the Triassic sandstone is only an occasional and not a predominant element in this part of the coast. The main mass of the Cumbrian promontory is Silurian, one of our principal mountain-builders. In fact on this promontory England's only real mountains are all situated in a compact family group, and it is rather typical of our perverse spirit to call this region the Lake District. But the name dates from the Renaissance period when lakes were far more fashionable than mountains.

However, the English mountains have a perfect frontage. If a West End window-dresser had arranged this part of the coast he could not have done it better. Our most important mountain is Scaw Fell which rises to 3,210 feet. It has a general south-west aspect, envisaging the mountains of Man across one half of the sea and the Mourne mountains in Ireland beyond the farther half. The mountain falls uninterruptedly to the coast, which is cut away as clean as with a knife for seven miles from St. Bees Head to the Duddon estuary. From the sea, on a fair day, the mountain is beheld rising towards the middle of this expanse of coast, receding as it rises and turning bluer as it recedes till a feather of cloud gives the right contrast to match the blue of the summit with that of the sea. Between these blues lies a coastal foreground of light gold from the sands and sandhills of that seven-mile stretch. The picture is completed by the Rivers Esk, Mite and Irt combining to form one break-out in the very midst of the golden bar. Here stands Ravenglas. It was a port in the days of the Romans, and small coastwise shipping can still make contact with the land there along its scored channel through the sands.

This great and glorious beach has but one formally organised resort for summer visitors, namely Seascale. That red glare on Skiddaw which warned the burghers of Carlisle of the coming of the Spanish Armada portended also for its neighbouring summits the arrival of a most unexpected denizen whose increment was from thenceforth to be one of the familiar local sights of Cumberland and Westmorland. For whereas the wrecks of that great fleet left names everywhere round our coasts, treasure-trove in a few places, and original patterns for Fair Isle jumpers in one, it gave the Lake District a flock of Spanish sheep now known as the Herdwicks.

Beyond the Duddon estuary, whose exquisite charms have been much marred by iron workings, comes a promontory known anciently as the district of Furness. I do not know whether the name is connected by an inverse process of derivation with the word *furnace*, but in historical significance the two are inseparable. From the earliest days of the settlement of holy men at Furness Abbey a source of wealth has been the iron (red haematite) which lay immediately below-ground in this district *plus* the trees of the forest which grew on the surface. By converting the latter into charcoal the necessary small furnaces (called of yore *bloomeries*) were made by which the iron ore was melted and converted to the use of man. The district, in fact, has had a parallel history with that of the Forest of Dean. The iron smelting trade was established here on traditional lines, and for this reason it has remained, although the immense blast furnaces that now redden the sky at night no longer use the local ore but that of the Rio Tinto mines from the land whence by a different accident, the Herdwick sheep came. In the eighteenth and nineteenth centuries when, with the Industrial Revolution, the first great vogue of metal castings came in, the centre of activity was not here but in Neath and Swansea. Still, it was from the families of the Furness ironfounders that many of the leaders as well as the workmen were recruited. One of the greatest of them, the ironmaster Wilkinson, was brought up to the art here but made his name and fortune with it in South Wales. He came back, however, to his native land to die and be laid, according to his wish, in an iron coffin. He it was who first thought of and made the iron ship. And it is a stroke of historic justice that the greatest English iron ship-repairing yards and armour-plating plant should be almost at the place of this man's home and tomb. I am thinking of Barrow-in-Furness. Barrow's neighbour, Ulverston, is perhaps the most powerful contributor of all those tawny beacon lights whose glow is seen from far over the eastern Irish Sea at night.

With Ulverston we have turned another corner and are in Morecambe Bay. This, in spite of industrial disfigurements on both sides, is still one of the most lovely inlets in all Britain. The mountain which dominates it is Black Coombe.

17 SILLOTH, CUMBERLAND: the Solway Firth, looking to Criffel Hill and Scotland

18 SUNSET OVER MORECAMBE BAY, LANCASHIRE

Its inward edge has no bay contour in the accepted sense but is fretted into the mouths of three estuaries, two of which are deeply indented, leading the salt water into the fresh under limestone crags and woods of birch and oak. Into one the waters of both Coniston and Windermere discharge. But the most beautiful is the inlet which goes up to Arnside and Milnethorp—the latter used to be the only port in Westmorland. I say *used to be* as I fancy that it is many years since a ship of any kind has been there, and the railway bridge would prevent one getting there now.

Morecambe Bay has one beauty of its sands and estuaries and another of its distant mountains. David Cox was among its best portrayers. The name Morecambe is Roman, probably adapted from a Brythonic word, for *mor* is still the Welsh for sea. The Roman version was Sinus Moricambe, and the only change that time has wrought on the word is that of substituting an *e* for an *i*. But elsewhere there is a Moricambe Bay with the Roman spelling intact. This latter is situated in the Upper Solway in the region of the great sea cattle-pastures near Silloth. It is a similar opening on a smaller scale, and the name is evidently intended to denote that it should be treated not as a bay but as a small sea with a sovereign dignity above bays and estuaries. One is reminded of the Morbihan in Brittany, and it is quite possible that in Roman times there was more of a barrier across the mouths of the entrances of these Moricambes. The present-day map shows clear indications that this was so.

But if Morecambe has had to have the commoner tag of "Bay" appended to it, it has succeeded in conserving the atmosphere, as well as the tradition, of a separate maritime entity. The Morecambe Bay fishermen with their specially evolved cutter-rigged smack called a *nobby* have been for generations famous throughout the three western seas of Britain. The boats were designed to stand the weather of the bay, for Morecambe gets the fullest force of the Irish Sea weather, there being 125 miles of open water between Heysham and Dublin. Even this length gives a short wave with steep sides and none of the generous length of the roller. This short, sharp sea encountering the shallows of Morecambe Bay makes ill weather for small shipping. But the Morecambe Bay nobbies would stand nearly anything.

Cranforth, the last of the blast-furnace towns, is on the

easternmost point of the bay. Here, too, the characteristic scenery of the west and north shores ends with a typical note in the limestone crag of Arnside Knot, with its broken peel tower. In passing, I think I should note that it is from the neighbouring village of Warton that the Washington family came. They were here before they moved to Sulgrave. Nothing is left of their old manor-house except the memory of the site, but the tower of the church was built by the family in the fourteenth century and bears the Washington arms, from whose heraldic device were derived the national symbols of the United States of America—the eagle, the stars and the stripes. From Bolton le Sands the tide recedes so far that it is possible to cross to the opposite coast of the bay, near Cartmel, a matter of about seven miles, which saves four times the distance in going round. Before the coming of the railway, this dangerous short cut was regularly undertaken by pedestrians, riders and even coaches. In earlier days a hermit guide had his cell at Bolton. But century by century the sea took a large toll of victims.

The shallows of Morecambe Bay are the foundations of a prosperous industry in shrimping, the wily crustacean being taken in a beam trawl dragged over the bottom by a nobby or other craft. There is an interesting contrast to the endless sandbanks, one of those local features which makes the bay such an individual sheet of water. This is Heysham Lake—not one of the recognised tarns of the Lake District. In fact it is invisible to the untrained eye at high tides in fair weather, for it is a sea lake.

We will return to it anon. First let us pause to lift eyes inland and see the River Lune, one of our most romantic salmon rivers, penetrating the Westmorland hills at the Tebay Gorge and pouring south by many a flashing fret and dark seductive pool till it passes under Smeaton's aqueduct (one of our neglected architectural masterpieces) and enters Lancaster with a truly dignified fluviate bearing, befitting its arrival at the seat of the famous and ancient duchy. Lancaster is a golden city, all built of the yellow millstone-grit, which follows the limestone down the coast in geological succession. And it is rather typical that Lancaster should be made of this material, for to the collier it is known as the *farewell rock*, meaning that when, in sinking a shaft, he strikes it, he expects to find no more coal. And, in fact, the awful squalor, the rural and social pollution which stamp the coalfields and industrial centres of Lancashire have not reached and affected the county town.

Lancaster's heyday as a port was when the first consignments of American cotton came over. It is said that the very first shipload to arrive was landed on the Lune estuary at a little place called Sunderland. This I had from an old sailor at Glasson Dock who told me, moreover, that a boll of cotton from this historic cargo had been planted on the spot, which later grew into a cotton bush and *still flourishes*. I have always meant to test this assertion by a personal visit, but the opportunity is yet to come. My informant went on to say that a negro was buried in a field close by the tree, under a plain slab of stone. A strangely appropriate conjunction when one considers how much the wealth of Lancashire in general, and Liverpool in particular, has been built on the unremembered services of the black man.

Heysham Lake abuts on the scour of the Lune through the sandbanks, and to this it owes its depth. It is three-quarters of a mile wide by four miles long, and as much as sixty feet deep in places. It rendered invaluable services in the old days of sail, for at the times between half-ebb and half-flood the surrounding sandbanks isolated it from both the force of waves and currents. In other words, it was as good a haven of refuge to any vessels caught by a south-westerly gale on one of the most dangerous parts of the coast as though it were protected by walls.

Heysham Lake served the age of steam in a different fashion. It enabled the Midland Railway to build a harbour from which they could compete for the Belfast trade against the steamers of the Lancashire and Yorkshire Railway that had long been running from Fleetwood. Accordingly they built a harbour at a convenient point on the Heysham Peninsula from which they could utilise the deep waters of the lake for making an offing. After the amalgamation of the railways Heysham replaced Fleetwood, which is now concerned entirely with the fishing industry. But there was probably a trade route to Ireland beginning in Heysham Lake a thousand years before Fleetwood was thought of as a port, for at old Heysham village there is a ruined church dedicated to St. Patrick.

Heysham has also replaced Morecambe as a port. The only marine activities there nowadays are sea-bathing and ship-breaking. Many a well-known sea favourite has crept here to lay her bones, including, I think, the *Adriatic*. As a seaside resort Morecambe is an old favourite with the busy towns of Lancashire and Yorkshire. It is still expanding and improving. Whatever its amenities, it has what no money could buy, an unsurpassed view of the Lakeland hills and the great inlet, which yields a different picture with every change of light from dawn to sunset.

The River Wyre enters Morecambe Bay just within its southern lip by an estuary which, owing to its mud-flats, used to be a famous resort of the wildfowler. At the mouth lies Fleetwood, a town half port, half resort, which is just a century old, and was born as the result of a railway scheme projected by Sir Peter Hesketh Fleetwood. Between here and the mouth of the Ribble there is an interval of low-lying coast which since the time of Doomsday has always been called the Fylde. For some reason it is a name which is only used locally and does not appear on maps. The Fylde is a territory just about as broad as it is long, extending inland from the sea to the hilly moorlands which are called the Fells and are the outposts of the immense tract known as the Yorkshire Moors of the West Riding. The name *Fells* carries us back to the time when, under Viking sovereignty, the Isle of Man ruled the waves of the Irish Sea. Although its name has been slighted by the map-makers, the Fylde is highly conscious of its individuality as a self-contained district, and as possessing a unique combination of bracing air and mild climate. Its best advertisement of the former is the long-maintained reputation of its seaside resort, Blackpool. Whatever criticisms people have to make about this, our most highly organised town of public entertainment, they always admit that Blackpool has bracing air.

It is a fact that local atmospheres do vary considerably. But this is not yet officially admitted by science, as the scientists cannot tell why they do so. And one has to go carefully in the matter as there is as wide a margin to the topic as there is to spiritualism. When I once asked a policeman at Port Stuart the reason why all the benches on the promenade had their backs to the sea he replied at once, "Ah, that would be because the sea air is so much stronger than it is at Port Rush that folks wouldn't be able to bear it if they sat facing." Blackpool has, at any rate, one obvious advantage in sea air above all other resorts in the Irish Sea. The prevailing west to south-westerly winds bring it not only a maximum of

19 LANCASHIRE AT WORK: Barrow-in-Furness

20 LANCASHIRE AT PLAY: Blackpool

21 THE MERSEY, AND LIVERPOOL WATER-FRONT

sea-borne air but also the most bouncing waves possible to raise, as it terminates the greatest open space of sea distance. The Fylde has two other seaside resorts, St. Anne's, which is on the turn of the land into the Ribble estuary, and Lytham, which is just within the mouth of that estuary.

The Ribble, which has an upbringing of great charm and picturesqueness among the Yorkshire moors and the Lancashire dales, is navigable as far as Preston. This Preston is distinguished from all others of that ilk by being called Proud Preston, though the map-makers have never accepted the name in full. The shipping of the port is chiefly concerned with the cotton-spinning industry of which Preston is an important centre.

South of the Ribble's mouth Liverpool Bay begins. This is a mere geographical expression like the Bay of Bengal, and means simply the south-east quarter of the Irish Sea, extending from the Ribble's mouth to Point Lynas in Anglesey. Geographically, the term is convenient, but scenically it is misleading, as the materials which make the coast vary from the alluvium of our present epoch to the oldest rocks discoverable in the world. The land lying between Ribble and Mersey is considerably flatter than the Fylde. It is the ideal pancake—without crumples; but unlike the Fylde, it has no proper name, on or off the map, the reason being that except on such eminences as Ormskirk, which rose from the chain of marshes and meres marking this sub-terrain in the Middle Ages, it has no history. It has a prehistory, though, for the dug-out canoes of lake-dwelling villagers are found.

The capital of this alluvial region is Southport, itself built not on alluvium but sand. In fact Southport is really more on the sand than on the sea, which is only just reached at the low tide by a pier 1,466 yards long. All the same, Southport has maintained an enviable reputation among seaside resorts for gaiety and healthfulness. The coast is all sandhills from here to the Mersey entrance.

The Mersey has a singularly inconspicuous youth. It comes from the Derbyshire hills, yet it only assumes importance when it reaches the Gap of Runcorn and passes through a gorge in the Cheshire sandstone. In the days when Halton Castle was garrisoned and Rocksavage Abbey sang praises to God this must indeed have been a romantic and picturesque situation. Even the most exquisite defilements of industrialism have not been able quite to ruin it. It is accompanied through the Gap by the Ship Canal which has given Manchester the privilege of calling itself a port in the Irish Sea.

The history of Liverpool is bound up with those flashes of benevolence which not infrequently radiated from King John where our seaports were concerned, especially those connected with the early Irish trade. His motives may have been inspired by money, or safety, or spite of the Cinque Ports, but the fact remains that he gave the first essential charter which put the Liverpudlians on their feet. In later history their expansion was prejudiced through a technical proviso which made them a "creek in the port of Chester." Shortly after this difficulty was overcome Liverpool had the satisfaction of seeing the port of its ancient rival silt up. The Irish trade that used to go to the Dee then began to come to the Mersey. But it was towards the Western Ocean and not the Irish Sea that her merchants had lifted their eyes. She began to compete with Bristol. Pioneers of the harbour-building era which set in in the middle of the eighteenth century report a strange lack of attention on the part of Bristol to the proper clearing of her bar and river-channel, owing to a long habit of supremacy. Of this Liverpool took full advantage and built its first dock. Then came the Industrial Revolution when business boomed in the North. Shipping crowded to the Mersey. And now, next to London, she has the largest range of enclosed docks in the kingdom. Her rival to-day is Southampton, which is successfully challenging the system of enclosed docks (that was Liverpool's pride) with a unique equipment of open quays where ships can come alongside at all states of the tide.

Unlike Southampton, Liverpool has got rid of all its old buildings. Although it has been established since the twelfth century, it is, architecturally, a seaport town of the mid-nineteenth. The future antiquarians of the thirtieth century would give much to see Liverpool as we see it to-day, for it is the throne of thrones from which Britannia ruled the waves in the period of Edward the Seventh.

Liverpool's docks, warehouses, floating stage and other port facilities are a grand spectacle. But the greatest thing of all about her that claims attention of the casual visitor is her entrance to the sea. The port authorities may frown over the d.fficult fairway, the long and costly revetment recently built out to the distant bar, the fleet of busy dredgers necessary for keeping the port open. But to the stranger, as he stands beside the bold brimming, tossing, tearing tideway of the Mersey and looks towards that pearly vagueness where ships outward bound dissolve and the homeward bound materialise, it will seem as if no other port in the world wears the mystery of the sea so near to its heart.

Across the water, on the Cheshire bank, is Birkenhead with its docks, and the Cammell-Laird shipyards. New Brighton is at the Mersey's mouth. Like Blackpool, it had its Eiffel Tower and its big wheel when these were fashionable as the nearest thing that men could get to flying. I remember going round in one of those big wheels with an old gentleman who swore that the Almighty would never permit men to fly as it would injure the prestige of the angels.

The southern limb of the Mersey which, in a short distance, becomes the northern limb of the Dee is called the Wirral Peninsula. It is a ridge of Triassic sandstone which rises to heathery heights, but is otherwise disposed pastorally in the typical placid, solid manner of the Cheshire dairy-farmer. The seaward end, at Meols, has extensive remains on the shore of a forest long since overwhelmed by the sea. The stumps of Meols have provided the universal text-book illustration wherever the subject of submarine forests is touched on. Hoylake and West Kirby are good examples of residential seaside places which are mercifully independent of the summer visitor trade.

Hoylake is on the open but shallow sea. West Kirby is just round the corner of the Dee estuary and faces Wales. If views could be judged on points, this one of Wales from Wirral, the honours of which (with slight differences) may be shared by Heswall, Neston and Parkgate, ought to rank among the prime views of Britain. Turner understood this, as I have said in the first chapter. To the tune of that view Handel put the finishing touches to his *Messiah*. This was, of course, coincidence as Handel was on his way to Ireland and was merely making the best of being weatherbound. But there always seems to me a true affinity between Handelian music and that grand and moving view across the Sands o' Dee.

Parkgate was an out-port of Chester, an effort to save the situation when the upper Dee began to silt up and the city of the Twentieth Legion ceased to think of itself as a seaport. For a short time Parkgate enjoyed some of the fame that had belonged to Holyhead and Beaumaris in being principal packet station for Ireland. But Rennie's new harbour at Holyhead, followed by Telford's new road to it from Shrewsbury, swung the Irish passenger trade back to Anglesey. The Sands o' Dee, together with Liverpool's newly won freedom, completed the ruin of Parkgate and closed Chester's long and honourable history as a seaport. But if you should visit the city and take a stroll on its ancient walls, notice a spur which has at the end of it a tower called the Water Tower. This spur is Chester's mediaeval pier. The ground about it is now not only dry but high. Yet if you search the wall of the spur you will find ring-bolts still in position for mooring ships to. And the sailors, who always take centuries to adjust their minds to new sea place-names, still never refer to the Dee by any other name than the Chester River.

Kingsley's poem relating to the sad tale of Mary and the cattle caught by the tide refers to the marshes of the upper estuary which, like those of the Solway, support large herds of store cattle. The beauties of this part of the estuary have been much impaired by reclamation schemes, and the whole of the rest of the Welsh coast towards the sea has been marred by the production of coal and lead and, to a lesser degree, the working of iron. For under the broad Dee sands lies a coalfield in which the miner works below the sea-floor. And the hills which rise up from the shore to grace the horizon with the Clwyd Range are made of limestone which yields rich veins of lead. They have been worked since the time when the Romans occupied Chester.

Of the Dee ports on the Welsh shore, Connah's Quay has sadly decayed for lack of sailing-ships. Mostyn is an ancient port which, if the scriptures in the Record Office and elsewhere were searched, would be found to have had an interesting, if not exciting, history. The general falling off of coastwise shipping has not affected this port, as Mostyn has an iron works. The entrance to the principal submarine colliery is at the Point of Air, where a quay brings shipping to within a few feet of the pit's mouth.

Between the Point of Air and its opposite number on the Cheshire shore lies Hilbere Island and its small sister, the Little Eye. The latter word, meaning *island*, has its local pronunciation Little *Ee*, which may be compared with the same word in the Thames, Chiswick Eyot, locally pronounced Chiswick *Eight*. Hilbere has associations with the pre-English

22 THE SANDS OF DEE: a Panorama of the great Estuary
at West Kirby, Cheshire

23 FLINT CASTLE: a medieval Stronghold commanding the
Mouth of the Dee

24 THE CLWYD ESTUARY AT RHYL, FLINTSHIRE

days, as it was long the home (and his cave is still shown) of that great Welsh saint of royal blood, Beuno. He was uncle to St. Winifred, whose well at Holywell, on the Welsh shore, is the nearest approach we have in Britain to the grotto of Lourdes. Hilbere is now a sanctuary for wildfowl.

The mouth of the Dee is an amazing complex of channels and sandbanks. When the marine topographer, whose advent I have prayed for in the first chapter, gets to work, he will find much of interest in the local names. Let me show a sample by quoting a whole passage from the *West Coast of England Pilot* (1922).

"The Welsh channel, which may be said to commence abreast of the Earwig bell-buoy, is about 5 miles in length to its eastern termination in Welshman gut, with a least breadth of 4 cables until close up to the gut, and with depths of from 6 to 12 fathoms over its central portion, but with many patches of less than 3 fathoms at either end. The Dee lightbuoy marks the fairway, and here the channel divides, the southern branch leading to Wild road and Mostyn deep; and the eastern by Welshman gut (said to be closed) leading into Hilbere swash and the northern Dee channel."

Gut is a fairly local term for a narrow channel which generally has a landward termination in a small creek. In the Severn Sea (or Bristol Channel) the same thing is called a *pill*. Swash, sometimes spelt swatch, is a narrow channel never terminating in the land. I fancy the name is of Viking origin. Elizabethan sailors used it freely. It occurs occasionally all round our coasts.

The Point of Air (probably a Norse name, too; there is one in the Isle of Man) is perhaps the most notable landmark in the whole of the Irish Sea. It stands for an astonishing break both in scenery and climate. You will notice the latter most markedly in the winter months if you are a railway passenger coming from the direction of Chester. That dismal uniform greyness which tinctures so much of the English daylight between October and April will cling to a train all the way from London. But as soon as the train rounds the Point of Air it will, ten to one, shake it off and find clear if not brilliant weather. Also from this point and Point Lynas there stretches an invisible line within which it is the rarest thing in the world for fogs to come. That heavy fog-bank, from which the Mersey shipping suffers so much inconvenience, assembles just beyond the line. It gathers alike on hot days and frosty days, and you can see the wall and roof of it from anywhere within the favoured zone of exemption, while the sun shines on you out of a blue sky. Rhyl, the first town beyond the Point of Air, holds the sunshine record for all Britain.[1]

From seaward, it is the Point of Air which marks the beginning of the Welsh mountains. To picture the dramatic suddenness of the change you must remember that the Southport coast was dead flat. The Wirral was a little higher, but seascapes are always flatter than landscapes. After that, there was the wide water-horizon of the Dee, only broken by the dot of Hilbere. The Point of Air is marked by an old lighthouse placed at the eastward end of a long line of yellow sandhills. Behind these the land has soared up, culminating in the great round boss of Moel Fammau, the mother of the Clwyd Range. This range is immediately followed by the Hiraethogs whose limestone front extends to the high cliffs of Old Colwyn, and is thrust out seaward to make one of the most remarkable promontories of all our coasts, the Great Orme, under which Llandudno lies. And this range is succeeded by a still higher one which contains Carnedd Dafydd, Carnedd Llewellyn (3,484 feet) and Glydr Fawr. Now the foreground of Anglesey slips into the picture, and behind it rises Snowdon, the most beautiful in outline as well as the highest in contour of all the mountains of England and Wales.

Between the Moel Fammau group and the Hiraethogs the fertile Vale of Clwyd comes down with its river of that name and opens to the sea at Rhyl. This, like all the seaside resorts on the north coast of Wales, Llandudno and Beaumaris excepted, dates only from 1849 when the Chester and Holyhead Railway opened up this fair seaboard to all and sundry. The place arose as an outpost of the very ancient seaport of Foryd which, in the days of the native Princes of Wales, was a maritime key of considerable importance. Since the War, shipping has practically ceased to come here. And now a new outpost has started to form a resort on the farther bank of the Clwyd. It is called Kinmel Bay, and is only just beginning to emerge from the bungaloid stage.

Prestatyn, however, and not Rhyl, is the first seaside resort which lies west of the Point of Air. Its origin was in

[1] Air Ministry. *Book of Normals of Meteorological Elements.* M.O. 236 (Sec. 1), p. 80.

a hill-foot village a little way from the sea. It has spread not only to the coast but right up into the Clwyd Hills. The beach of sand which reaches from the Point of Air to Kinmel Bay is replaced by one of white shingle confronting Abergele, with a sandy interval at Llandulas. Then follow the limestone cliffs of Old Colwyn, and, afterwards, the great sandy sweep of Colwyn Bay. The natural shapes of Colwyn Bay and of the countryside behind it are as lovely as anyone could wish for. The town itself is a great improvement on Rhyl, though it leaves much to be desired. But it has a certain lively sparkle of Continental vivacity which makes up for much. It extends as far as the ancient village of Llandrillo-yn-Rhos (Anglicised into Rhos-on-Sea), which must have been a charming place before the boarding-houses came. You may still discover the ruins of its fish-weir built by two Cistercian monks of Aberconway.

Llandudno Bay is surely the most lovely crescent into which the Irish Sea finds its way, and the Great Orme's Head gives it a most impressive finish. But the town, which is Wales's most pretentious seaside resort, lacks the frank (if vulgar) gaiety of Colwyn Bay. Beyond the Orme, the Conway River comes down and divides Tame Wales from Wild Wales, and the division is a positive one both as regards the mountains and also as regards accommodation for tourists. By which I mean that there is a cessation of big hotels, the boarding-houses are less pretentious, and things are, in general, more simple and more truly near the homely spirit of the Welsh people. The last resort in Tame Wales is Deganwy. It lies within the estuary of the Conway. Its name has resounded loudly in the long contests of history between the Welsh and the English, but its appearance is completely modern.

The estuary of the Conway has a grand debouchment between mountainous banks which, on the western side, are clothed with woodland. It expands into a wide lagoon, and then goes out to sea through a narrow channel amid sand-hills. At the point where, on the left bank, the mountain breaks off into a lower hill forefoot the town of Conway stands. Until the time of the Conquest of Wales in 1283, a Cistercian monastery stood here, and the beauty of its site, even for a Cistercian foundation, must have been unrivalled in Britain. Edward the First had it decorously moved higher up the river, leaving, however, its conventual church to serve as the parish church of the new town which he at once set about building. His walls still surround the town as it stands to-day, and they are among our most beautiful and interesting relics of the Middle Ages. Within the south-east angle of the walls stands the great castle whose ruin is more picturesque than any other castle in Wales unless you except Cerrig Cenin). Before the invention of the diving-dress there used to be an active industry at Conway of pearl fishing—there are Conway pearls in the British crown—an interesting relic of the industry remains on the sandhills near by, still called by the sailors Mussel Hill. It is a dark mound of shells.

To those who are not content to survey the mere print of history but would examine the page on which it is written, this mountain barrier which rests on the left bank of the Conway must seem a fundamental bedrock. Indeed it would not be difficult to show that this accident, geologically dating from the Cambrian Period, could be traced as a definite moulding force throughout the whole history of Britain, not to mention the whole history of the Irish Sea. These solid, lofty and ancient mountains are quick to set their seal on the coast by imposing on it an immense headland of diorite in the shape of Penmaenmawr. It is still a formidable mass. But its true majestic proportions which formerly marked it as one of the most outstanding features of all our coastwise scenery must be sought in the old engraving of the mid-nineteenth century. For the quarryman has now marred it beyond recognition. When will our rural reformers realise that the quarryman is an infinitely greater danger to the beauties of our landscape than the speculative builder? The works of the latter can be removed at any time by a few slabs of gun-cotton. Nay, time will do it for nothing, and before so very long. But the damage done by the quarrymen can never be repaired or restored.

Penmaenmawr and Llanfairfechan cater for visitors who wish to take their pleasure equally by the sea and among the mountains. For an easy access to the hills, Penmaenmawr is rather too steep-to, as the sailors would say. But from Llanfairfechan you can get straight into the heart of the Carneddau Mountains, carry your rod up to the mountain lakes, or take a boat and dabble in sea-fishing in the plaice-haunted channels of the entrance to the Menai Straits. There are two more swatches here. And it is said that in one, the

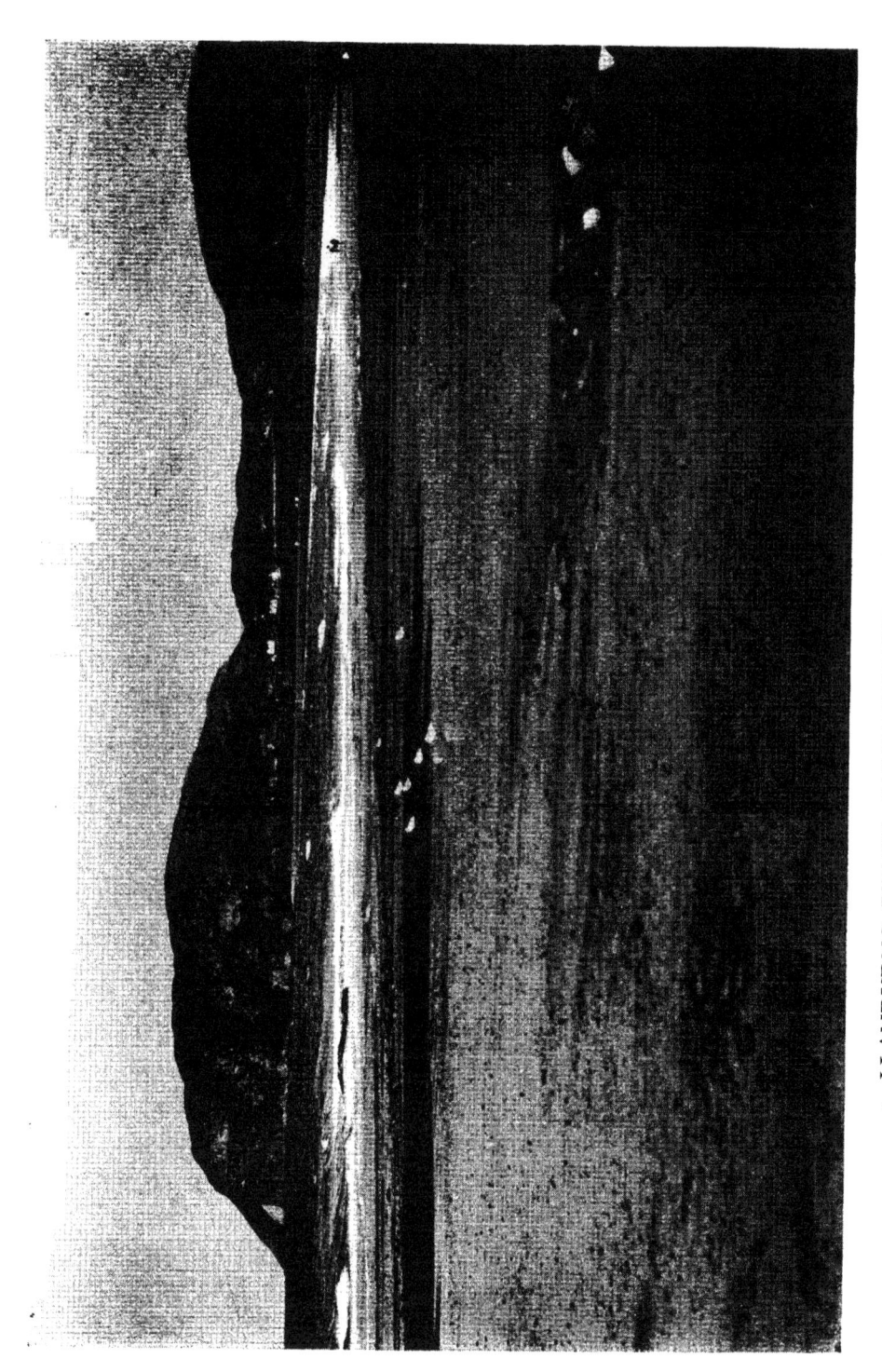

5 LLANDUDNO BEACH, CARNARVONSHIRE, looking to the Little Orme

26 LLANDUDNO, CARNARVONSHIRE : the West Bay

Midlake Swatch, some ship of the fifteenth century, carrying alabaster tombs from the Nottingham workshops on the far Trent to bereaved notables in these parts, foundered in a storm. And I have always liked to picture what the impression on a Celtic sailor or fisherman would be (especially were he a little primed) if, through that transparent water in the shallows, he were to catch sight of a white alabaster face peering up at him from the bottom, and, moreover, catch a glimpse of hands folded in prayer.

But the wrecks on these Lavan Sands have been by the thousand, and little ever appears to tell the tales of the sea. As a matter of fact, over most of these sands the sea has only rolled since some date in the fifth or sixth century. There is a lost land here well remembered in legend. Its chief city was Llys Helig, whose relics the author of *The Evolution of a Coastline* has professed to have traced just outside the Conway bar.

The wide entrance of the Menai Straits makes a quick closure at Bangor, one of the ancient cathedral cities of Wales—there were only four until the recent Welsh Disestablishment Bill went through. From here to Carnarvon, where the strait joins St. George's Channel, the wild sea water is tamed into a river treatment, passing along the verges of rich pasture land and hay meadows, and under the shadow of trees whose habit is not to grow near the seashore. It repays these attentions by rendering richer colour to the landscape which it divides than fresh water could do. The rapids of its tidal streams and the edges of its Caribbean whirlpools give off a dazzling foam; islands rise from the green flood which afford only just enough room for a cottage and its garden, while the seaweed and briny exhalations add their bouquets to those of wild and garden flowers. Across this strait hangs Telford's great suspension bridge which I have elsewhere extolled as the most truly aesthetic work of architecture or engineering that has been put up since the Middle Ages. I must admit I do envy the Marquess of Anglesey who can live in a house which has a quay at the bottom of the front lawn where he might keep a yacht as large as a cross-Channel packet tied up, and go to sea in her at almost any state of the tide.

I could easily linger for two or three pages over the details of this lovely strait. But as far as the Five Seas are concerned it is a no-man's-water. We must hie us back to the eastern entrance which can legitimately be said to be an inlet of the Irish Sea. Bangor has a boat-building yard chiefly given over to the construction and laying-up of yachts. It also has a pier with a floating stage at the end and used to possess a busy town quay, but this is given up. In addition, it has quite a large privately owned harbour, the property of Lord Penrhyn, who owns the famous slate quarry at Bethesda. If you should fall in with a steamer with a Welsh dragon on the funnel she will hail from this harbour, called Port Penrhyn.

Bangor is "a creek in the port of Beaumaris," to use the language of His Majesty's Customs, although Beaumaris has no longer any shipping facilities except a short pier. This place, like Conway, is a military plantation of Edward the First. Its castle remains in a very fair state of preservation, but its town walls are gone. In the early days of the Welsh seaside resorts Beaumaris was very popular. One wonders why it has failed to keep touch with the fickle tourist, for it has so much to recommend it. It is as "old world" as one could wish, is sheltered from the south-westerly gales, commands the most magnificent view to be had anywhere of the mountains and the Menai Straits, and has a fine old Georgian hotel. Besides, there is something rich and mellow about it which is missing in the majority of the Welsh seaside resorts.

But Beaumaris is not at all typical of Anglesey. It was planted with Anglo-Norman settlers in the thirteenth century. And oddly enough, from totally different causes the same blood was reinfused into it in the early nineteenth century when families of fishermen from the Channel Islands came to settle there. So among the Williamses and Robertses you find Perchards and Galachens. Also the country around is well wooded, which is quite contrary to the rule in Anglesey. I even know of a well-grown eucalyptus-tree there.

Elsewhere in Anglesey the two things are obvious on all hands—it is very Welsh and very wind-swept. From a motor-car the island appears flat and uneventful. But you can never see Anglesey from a car. You must walk or cycle. If you try the latter you will find the country very hilly. But you will be able to study the small features, and it is these which count in Anglesey: the small fields with their rough grey stone walls and their black cattle, the little whitewashed farms, the flourishes of gorse and bracken, small pastures that are green and rich in spite of the frequent outcropping of grey, twisted rock. But it is the brilliancy and the taste of the air which unifies the scene into something exceptionally beautiful and individual. All hilltops give two superb horizons, one of the sea and another of the mountains.

Anglesey has an island at each of her four corners, three of which are dedicated to saints, and I fancy the fourth was, only history has not kept touch with him. The corner near Beaumaris has Puffin Island, a name of the late English period which replaces an interesting Scandinavian forebear which was *Priestholm*. But the Welsh name, still used, goes back to the Age of the Saints, which was the sixth century. By interpretation it is Seiriol's Isle. If the Great Orme be likened to one Herculean pillar at the mouth of the Menai Straits, this is the other. It is also made of limestone and, like the Orme, stands high out of the sea, having the still intact tower of a Norman church on the middle of its lofty skyline. There is a deep water passage between the island and the mainland called the Penmon Sound, through which the tide-rip roars and tosses as it does between the Isle of Man and the Calf.

From Penmon Point to Point Lynas there are many charming sandy bays, the largest of which is Redwharf and the most dainty Dulas Bay, where the red Devonian rock suddenly makes its appearance on the coast. Benllech and Tyn-y-Gongel have resorts—as this term is understood by the Anglesey builder, and Moelfra, which used to be one of the quaintest little fishing villages in Wales, is now rapidly "developing."

Point Lynas is the outpost of the Liverpool pilotage and the pilot steamer cruises there in the deep water right under the land, day and night, dodging the point for shelter. I have not counted Point Lynas as one of the four corners of Anglesey, but the coast takes a definite turn here and assumes a different front towards the sea. It is now rocky and cliff-bound with many beautiful small headlands and cleft inlets, but with no sand except at Bull Bay and Cemaes. There is a remarkable little town here with a rock harbour which, instead of getting less picturesque every year, like other Anglesey coast places, gets more so. In the hill above the town an immensely rich find of copper was made in the early nineteenth century. In fact it proved to be the richest mass of copper known at that time. An important mining industry quickly grew out of it and several subsidiary industries formed round that nucleus. Amongst these was a manufactory of twist and shag tobacco for the consumption of the workmen. So busy were things at Amlwch that the harbour would not hold the shipping that was needed. There was, in fact, a queue, and it had to ride at anchor in Holyhead Harbour and wait turns. Then, just a few years before the War, the supply of ore gave out.

One by one the subsidiary industries closed down till there was nothing left except the making of tobacco. There were three such factories. In a few years Amlwch became as derelict as an old battlefield. Its depressing appearance kept summer visitors away. It was the twist and the shag which saved the situation in the end. These factories never closed down, as every Anglesey farmer and rabbit-catcher had come to swear by *bacco Amlwch*. By now, the worst ruins have been cleared away and the rest have got to look as picturesque as the relics of mediaeval feudalism. And the rejuvenated cottages by the old rock harbour are coming into favour—in short, the place is being "discovered."

Carmel Head is the north-west corner of Anglesey—a sailor's name, displacing the Welsh Trewin-y-Gader. It is a serene place, and I could never compare it with any other headland. The smooth moors sloping down from Garn Hill rest on a massive cliff pedestal of Ordovician rock. The sea to north and west stretch away bounded by one horizon. Two miles out lie the Skerry Islands, whose light is now the most powerful round the British coast. The right to illuminate these islands with a coal-fire beacon and charge dues therefor had been in the hands of a private family from the time of Queen Anne until the middle of the last century, and Trinity House had to pay just under half a million pounds to secure the site for their lighthouse. The water here is immensely deep right at the cliff's foot, and in colour a strong blue. But it is never still as the tide pours by, filling and emptying the Irish Sea. From this corner, though you may not see the coasts of Ireland, the Isle of Man, and Cumberland, you will generally see the high piled clouds that are gathered above them gilded with the lustre of sea distance, and stock-still.

Anglesey's west coast has low cliffs and many sandy bays. A strait (for which I have never heard a name) separates the main island from the largest of her islets. The Welsh name

27, 28 THE PUFFIN AND ITS FAVOURITE HOME: Puffin
Island off Penmon Point, Anglesey

29 TELFORD'S GREAT SUSPENSION BRIDGE TO ANGLESEY OVER THE MENAI STRAITS

commemorates the name of the sixth-century saint who founded a monastery (of the Celtic sort) here. For which reason this island is called in English Holy Island, and the town which has gathered about the monastery under the mountain headland is called Holyhead.

Holyhead has been a port for Ireland from times out of mind. As early as the seventeenth century there was a mail contract with an Irish ship-owner. This became unsatisfactory, and after Telford's new road was built and Rennie's harbour, the Post Office themselves became ship-owners and operated a fleet of brigs under the command of retired naval officers. Incidentally, *a propos* of what I said earlier on about the nomenclature of submarine topography, a captain of one of these Post Office brigs, named Skinner, was washed overboard and drowned. He was a popular man both in the town and with the sailors. The former have commemorated him by a monument. But the sailors have given his name to a reef of rocks—the Skinners. And this memorial is likely to outlast the tribute of the townsmen.

In the late 'forties of the last century the attention of all Britain must have been concentrated on Holyhead. The Government had decided to make it a haven of refuge, that is a port where vessels weatherbound by dangerous or contrary winds, at this junction of the two seas, could put in and lie safely at anchor without having to pay harbour dues. In connection with this, immense works were being put in hand. At the same time the railway was advancing. It had been a matter of speculation as to whether Holyhead should remain the mail packet station or whether this should be transferred to Porth Din Llein in Carnarvon Bay, a proposition which would save building another bridge over the Menai Straits. The proposal to build a harbour of refuge and combine it with improved facilities for the Irish mail service decided the authorities to retain Holyhead. A third piece of excitement was that the Post Office had made up their minds to cease from being ship-owners and to put the mail contract out to private enterprise again. The successful applicants for this were the City of Dublin Steam Packet Company.

Nothing turned out as was expected. The Government built one-half of the projected harbour and then changed the scheme. Instead of building the second wall they decided to carry on the first wall so as to turn the whole of Holyhead Bay into a harbour. This entailed that change of direction which makes Holyhead breakwater look like a snake and which carried it seaward for the prodigious length of a mile and a half. The protection afforded to the mail packet station based on Rennie's old harbour was, of course, magnificent. Beyond that, Holyhead breakwater has proved to be a failure, as in bad weather there is too great a "range" for large ships to lie safely there, and half the harbour has a foul bottom. Holyhead's future proved to lie in the old inner harbour. The mail contract having been given to an Irish shipping company, the railway decided to run their own steamers. When all these schemes had first been mooted, the railway had been only a comparatively small concern whose termini were respectively Holyhead and Chester. It had hardly run its first train, however, before an amalgamation was arranged with lines which had been growing to meet each other for the last ten years, between Chester and London. The new system was known as the London and North Western Railway. It grew to be the most highly efficient organisation of its kind in the world. The only one of its major ambitions which was never realised was the securing of the Irish mail contract. The City of Dublin Steam Packet Company proved equal to its powerful opponent. It held on grimly to its mail pier at Holyhead in spite of everything the railway could do—and the railway tried means fair and also foul. It was the keenness of this immortal contest which by 1914 had established the Holyhead–Kingstown service as the most efficient cross-Channel service in the world, with the fastest, largest, and most comfortable ships.

The story is one of the most romantic and dramatic in commercial annals, for the two protagonists finally died in each other's embrace. Our grandest and most individual railway suddenly lost itself in an extinction more ignoble than death, namely absorption into an unholy admixture with its older and inferior competitors—it perished by amalgamation. At the same time the City of Dublin Steam Packet Company went out of business. True, the railway now holds the mail contract, but it is not the same railway. The ferry does not run to the same Ireland. The only things which are still the same in spirit are the train, Irish Mail, incidentally the oldest train in the world, and the Holyhead packet, an amazing combination of efficiency brought to ripeness by that titanic struggle in the past.

The Mediterranean has hardly a finer ceremonial entrance

than the Irish Sea. It would be a mistake to compare Gibraltar and Holyhead Mountain too closely. There is, however, one identical thing about the two, lying in the idea which each inspires: an idea which is one of the most inspiring if indefinable in all geography—the parting of the ways of two seas. By 365 steps you may walk down the face of the great precipice of the south front of Holyhead Mountain, which is made of the oldest known rock in the world, pre-Cambrian quartzite. By a little suspension bridge you may cross over the surging sea gulley and gain the South Stack Island. Here you will feel that you have arrived at a place which has a force of character quite dynamic. From the sound of the sea to the aspect of the horizon it gives the impression of a wedge between destinations, a place of passage. For the birds who crowd every inch of ledge on the furrowed cliff, as well as for sailors all over the world, Holyhead is one of the principal landfalls and departures.

CHAPTER III

ST. GEORGE'S CHANNEL

Even to him who looks at the sea from the land only, St. George's Channel will have certain marked differences from the Irish Sea, for the waves are longer and the water is clearer. The Atlantic rollers, running under a sou'westerly wind, bowl in through the southern entrance, whose gate-posts are Carnsore Point in Ireland and St. David's Head in Wales, except within the crook of Tremadoc Bay where, from various causes, the bulk of the sand lies, the water is deeper than in the Irish Sea and the shoreward tides less turbulent, which facts account for the greater general clearness and blueness of the water.

In the public consciousness St. George's Channel has much less identity as a separate sea than the Irish Sea. There are two fairly obvious reasons for this. The first is that there is not a single considerable port either of origin or of call within its coasts. Secondly, St. George's Channel is a name scarcely mentioned except on the map. It is generally thought of in coastwise terms relating exclusively to this side where its two main features are Carnarvon Bay and Cardigan Bay.

The Irish coast, which extends from Bray Head to Carnsore Point, is singularly uneventful except that at the northern end it has a splendid background in the Wicklow Mountains. But it has one feature of peculiar interest, namely, a barrier-reef. This lies a few miles off the shore and keeps on cropping up under the water in a direction exactly parallel to the land. The three principal outcrops are the Kish Bank, the Arklow Bank and the Blackwater Bank.

The ports are Greystones, Wicklow, Arklow, Courtown Harbour and Wexford, with its outport of Rosslare Harbour to which the Great Western steamers from Fishguard ply. With the exception of the last named, all these ports may be described as decayed, though perhaps that is too hard a description of Courtown Harbour which is a charming little place.

On the Welsh shore, St. George's Channel is a sea of mountains and magnificent estuaries, and, where these fail, of grand cliffs.

30 CARNARVON FROM ANGLESEY

31 ANGLESEY, with a distant View of the Britannia Bridge, Menai Straits, and the Church of St. Tisilio

We may take it in detail, beginning where we left off at the end of the last chapter, by the South Stack, off Holyhead Mountain. There are two well-known places on the south coast of Holy Island where the summer visitor takes his pleasure, Trearddur Bay and Rhoscolyn. The former is built up, though with more regard to appearances than most Welsh resorts. It has, in fact, three hotels and a thriving preparatory school. Rhoscolyn has no accommodation to speak of (mercifully). It is one of the most lovely little bays in Wales, and its wide reputation evidently rests on memories of picnics. In Anglesey, near the strait which separates it from Holy Island, is Rhosneigr. It has a fine beach and is Anglesey's most pretentious seaside resort. In bygone days it was celebrated for its wreckers and its smugglers. There are many tales of how these fellows hid both their booty and their victims in the sandhills—a wonderfully safe *cache* if you had the correct marks for rediscovering the goods.

Until the post-war days of multi-motordom no visitors except a few initiated spirits troubled the rest of this part of the Anglesey coast. Its delights consist of low cliffs of grey lichened rock that can be easily climbed at nearly all points, headlands surmounted by heather and a springy sward that turns azure with squills in early spring, and later, is jewelled with the rubies of thrift and centaury. These headlands enclose charming little sandy bays. There is a church in the middle of one, for the sea has encroached and left it, with part of its graveyard, an island. Near this church is Aberffraw, a tiny village now, but once the capital of all Wales, when the princes of Gwynedd were the paramount rulers. Llewelyn ap Gruffydd, the last of the native princes, had his chief palace here. Nothing remains of it now, for most of the buildings in Wales at that time were of wood.

At the south-west corner is the great warren of Newborough which assumes in its midst all the proportions of an African desert. Off its coast lies Llanddwyn Island, which is perhaps the chief gem of Anglesey scenery, or was until the coming of the motor-boat, for the Newborough Warren defended it handsomely from the motorist. Its Celtic monastery was founded by Dwynwen, the patron saint of lovers. There are no remains of this, but there is a beautiful fragment of a sixteenth-century church left standing without any history to match it. Llanddwyn is the pilot station for Carnarvon and this end of the Menai Straits. Here you may find a sheltered sandy bay in any wind that blows. Some of the rocks on its foreshore are blood-red jasper, others are pink marble veined with white. It looks across the inner sweep of Carnarvon Bay to a panorama which extends from the greater mountains to the striking blue peaks of the Rivals in the Peninsula of Lleyn.

Carnarvon with its splendid castle and mediaeval waterfront lies just within the southern mouth of the Menai Straits, confronting a wide lagoon bitten into the waste of the Newborough Warren. Shipping can still enter the old harbour, which is in a pool at the mouth of the Seiont River under the high walls of the castle. The new harbour with its docks lies beyond the walled town and was a "forest of masts" within my recollection. It is rare to see one ship there now. An interesting feature at the narrow entrance to the straits is Bellan Fort, a private gift to the Government at the time of the Napoleonic Wars. It is equipped with a small dockyard for building and repairing ships of war such as Nelson's navy used.

The Peninsula of Lleyn, which is the northernmost and smallest of the three characteristic limbs of west Britain that are flung at a south-westerly angle into the sea, is always spoken of in Wales as a region apart. And the levelling influences of the present day have not removed that peculiar charm of detachment which characterise both the people and their wild scenery. Its small mountains, its moors, its tremendous cliffs, and its interspaces of rich fertility all seem to be a special creation made for the benefit of Lleyn. On the north coast, Nevin is the only place which thinks of itself as being a seaside resort. It is a very Welsh village with many apartment houses added unto it.

On the south coast, which ceases to be cliff-bound at Llanbedrog headland east of Abersoch (a tourist centre rather more than just in the bud), there is Pwllheli which has had a long but never exciting or brilliant association with the visitor trade, and Criccieth which has always done things on a more ambitious scale. Nature and Edward the First have done for Criccieth what, so far, the Welsh building contractor has been unable to undo. It would be hard to spoil that splendid bay and distant view and the picturesque ruin of the castle. In addition, Criccieth has a certain charm of true Welsh homeliness that doesn't always get through at these points of all-English contact.

But it is the end of Lleyn peninsula which is the most stirring

32 NEVIN BAY, CARNARVONSHIRE Looking to the Rival Hills

33 THE HARBOUR AT PORTMADOC, CARNARVONSHIRE, looking to Cnight Mountain

to the imagination. Here lies Aberdaron, the virtual capital of the district, at the bottom of a rock-girt ravine which cannot be improved on in Cornwall. Beyond that sheltered bay is the tearing race of the Sound of Bardsey (Braich-y-Pwll in the vulgar tongue) which runs between the peninsula and the most noted of Welsh saintly islands, Bardsey. Bardsey Island (Ynys Enlli) had a truly historic monastery which was of the Celtic and not the Roman order, but of it there are no remains; only a tower of its Benedictine successor still stands. The small population is ruled by a farmer-king whose style and authority may be traced back to the abbots of Bardsey. St. Tudwal's Isle and its fellow lie off Abersoch, but their history is obscure.

The base of Lleyn is rooted into the highest of the Welsh mountains and the south shore runs up into a creek which is dominated by Snowdon, Moelwyn and Cnight. Here stands a centre of early nineteenth-century enterprise in Wales, the embankment, across the Glaslyn estuary, the land reclamations and the slate port all created by one Madox whose name, having been gallicised, is commemorated in Port Madoc. The economic excuse for impairing the grandest prospect of mountain and estuary in all Britain was the getting and dispatching of slate. The usual accretion of adjuncts followed, the principal being an iron foundry. Port Madoc had a long reign of prosperity which lasted until the outbreak of the War. Now its harbour is empty, for the slate industry flourishes only at the mountain end of the famous Blaenau Ffestiniog–Portmadoc Narrow Gauge Railway, and the finished product goes off by train. The other industries are gone. Once more it is the natural grandeur of this extreme north corner of Cardigan Bay which asserts itself, and in a creek withdrawn from the unsightliness of Portmadoc's decline a new seaside resort on entirely original lines has been started. This is Portmerion. The layout is in pseudo Italian style in which bright colour plays a leading part. I could have wished that such a longed-for innovation had been a serious attempt to draw the genius of Wales, which is only dormant for lack of development, and to have made a resort that breathed, both in catering and architecture, the native spirit of hospitality and homeliness.

The easternmost inlet leading out of the wide estuary on which Portmadoc is situated is narrow and has for its banks the mountain-sides themselves. This fine glimpse is denied to the traveller by road though the railway passenger sees it as he crosses the water on the low trestle bridge. Here a wide expanse of sandhills begins, filling up the corner of Harlech Bay. They appear to have accumulated since the Middle Ages when free access to the great rock on which Harlech Castle stands formed part of its defensive scheme. The rock of Harlech with its grand ruin is one of the most striking accents in Welsh coastal scenery. The close-up view is as grim and forbidding as its military creator desired, but seen from anywhere on those wide sea horizons which it commands it appears as a jewel on the bosom of the hills.

The coast from Harlech to Barmouth is the seaboard of the ancient domain of Ardudwy (the *u* is pronounced as an *i*). It was part of the Kingdom of Dwynedd but had a separate pride of its own. To the visitor it appears distinctive enough by reason of its mountain-shapes, which are loaf-like (I am thinking of bread, not sugar) rather than peaked. The chief of the range is Rhinog Fawr which reaches a height of 2,362 feet. The valleys of the Rhinog Range are well wooded and luxuriant in ferns and mosses, the sides of the mountains deeply quilted with heather and bilberry. In every available hollow there is a lake. They have indeed a sparkling scenery which seems to have bred a gayer and less solemn spirit in the people than the steeper hill country of the north. But the sea coast is low everywhere, except at Harlech, with an endless foreshore of sand. All the villages have an eye to summer visitors.

There is a lost land off this coast whose relics may still be traced in the great reefs of Sarn Badrig (St. Patrick's Causeway) Sarn Bwch and the Gynfelin Patches. Our nautical guide, the Pilot, states: "At the eastern end of Outer patch are three large stones which formerly uncovered at low water springs, but are now always covered. One of these, measuring from 15 to 20 feet across, had the appearance of being a mass of ruin." Peacock, one of the few English authors who have used the rich local colour of Wales as a background for fiction, wrote his *Trials of Elphin* about this lost land.

Barmouth is a corruption of the Welsh word Abermaw, which means the mouth of the Maw and not of the Bar. It has its little harbour as well as its seaside resort. The entrance to Barmouth estuary, seen from seaward, or from the railway bridge which goes right across it, is the most picturesque in Wales. The striking precipitous table-mountain of Cader Idris stands on the southern shore, while below this serene figure there appears a perfect *cheval de frise* of hill and rock contours.

4 BORTH-Y-GEST, CARNARVONSHIRE looking down on Carreg-Cnwe Ġe

35 HARLECH: the Town and Castle from the Sand Hills

Between this estuary and the third of this striking trio which carries the waters of the Dovey to the sea there is a precipitous piece of coast along which the railway has to go on a hewn ledge. The remainder of the coast is flat. On that part stands Towyn, which is an ancient village with a modern sea-front—the two are separated by a mile of no-man's-land.

The Dovey estuary, which has Aberdovey on the northern bank of its entrance, is the boundary between North and South Wales. Its scenery typifies the difference of the two halves of the Principality in a very striking manner. For you may read the sterner, wilder, more inflexible character of the northern people in the rugged peaks of Cader Idris and the Arans which tower above the right bank of the Dovey, while in the flowing winsome contours of Plynlimmon, on the left bank of the river, you may see a clue to the temperament of the South Walians. This is a very remarkable example of environment making man and of geology making history, for the Dovey is a positive break in the scenery of Wales. The Plynlimmon group which extends in the background of the coastline almost as far as St. David's Head is composed of Silurian shales. Inland this forms deep, well-wooded ravines where the rivers run, and on the coast it makes fine cliffs blueish in colour, veined with white, shining quartz, and cut to jagged and interesting silhouettes.

Borth, like Llanfairfechan and Towyn, is an old village which has espoused a modern waterfront. The coast is low with a wide marsh between the shoreline and the mountains. But Aberystwyth is in close touch with the hills, and the two rivers, Ystwith and Rheidol, which join actually under the castle walls and form the harbour are nobly born from the Mother of Wye and Severn and have a singularly beautiful youth in the rich valleys of Plynlimmon. They come down to the town with all the brimming freshness of mountain streams. This sea-front is a good one from which to observe big waves, and the sou'wester sends some fine specimens along.

Aberayron has charms for the lover of eighteenth-century Wales but no great attractions for the summer visitor. None of this coast is particularly interesting till you get down to New Quay, by which time the cliff scenery of Cardiganshire has well begun. New Quay Bay is hooked in behind New Quay Head where it lies sheltered from all winds except those which come from the north and north-west. With this situation it commands a unique view of the coast as far as Bardsey

Island and nearly all the principal summits of Wales, in which you may include the mountains Plynlimmon, Cader Idris, the Arans, the Rhinogs, Snowdon himself, and many of his neighbours.

Without disparagement, New Quay would be better described as a resort of seamen than a seaside resort. They say there is hardly a house in the place where you cannot find a retired sea captain or the relative of one. For my part, if I were to seek the seaside for a holiday (for the sake of the sea) I would greatly prefer to find myself among a community of mariners than of pierrot-loving landsmen and temporary beach-combers.

Between New Quay and Cardigan town there lies one of the few remaining strips of our coastline that has not yet been "developed." Unfortunately that dread process is beginning. If the local owners of property at the four places I am about to mention had foresight they would immediately get together and formulate some plan which would insure that their amenities in scenery were damaged to a minimum degree by the expansions which are bound to come. If they resolved on a scheme that was really exemplary they would never need to waste money on advertising. If any reader should be allured to these parts by my advice I hope they will deliver my message personally.

The places are Aberporth, Traeth Saeth, Llangranog and Penbryn. I have mentioned them not in geographical but in developmental order. Aberporth is perhaps beyond the reformer's aid. Traeth Saeth is only in the railway-carriage bungalow stage. It is a beautiful green hollow let softly into a gap in the blue-grey cliffs. Just above its crisp, clean apron of sand stands its lime-kiln (a relic of the agricultural revival) looking as picturesque as a ruined mediaeval tower, but just because it isn't one, in danger of being improved away. You may gauge the rusticity of the place when I tell you that when I was there last year I found a Dartford Warbler's nest in the thick of the bungaloid growth.

Llangranog, has a tradition of seafaring and even shipbuilding; and established places are always harder to spoil. But the charabanceers from the industrial areas of South Wales have found it out for Sunday outings. It has an ideal little strand with cleft and towering rocks. Penbryn is an untouched and unbuilt-on bay, the most beautiful of all of them. I pray to God that the National Trust gets hold of it before the speculative builder. Its beautiful little churchyard, tucked into a hollow behind the bay, holds the ashes of Allan Raine the authoress.

The town and port of Cardigan lies three miles up the estuary of the Teifi, a river whose upper stretches are noted for their gorge scenery. The mouth of the estuary has fine sands which attract bathers in the summer. The town was one of the earliest Norman settlements in Wales and, coming from the north, which had no Norman Period, one notices the distinct influence on architecture and building in general. From here round the coast to the Gloucester border, this influence prevails in town and village.

The finest cliffs in St. George's Channel are probably those between Cardigan and Fishguard. They include the great wedge-shaped headland of Dinas which achieves 460 feet. But they are not well known as there are no interludes of sand except at Newport. This place lies a few miles within the Pembrokeshire border which is defined by the lower Teify River and its estuary as at Cardigan town. I fancy that of all our counties Pembrokeshire could claim the most interesting variety of cliff scenery, if the islands off St. Anne's Head were included. Newport has the distinction of retaining closer links with the Norman system of government by the marcher baron than any other place. It is so well rooted in the past that two of its inns still brew their own beer. The sands are extensive, yet the place is so far unspoilt by make-shifts for summer visitors.

Fishguard has two fine natural harbours, the one right under the cliff on which the little town stands has been used since the Viking settlers gave it its peculiar name. The Great Western Railway uses the adjacent bay of Goodwick for the departure of its cross-channel service to Ireland. A breakwater puts the last touch to the splendid shelter afforded to Fishguard Bay by the bluff promontory called Pencaer whose corner-stone is Strumble Head. Incidentally, Goodwick Bay has a fine sandy beach.

This coast has a peculiarly interesting hinterland. It is even richer than Cornwall in wild areas where the remains of pre-historic man remain untouched. But the wildness itself is of no ordinary kind. Prescelly Mountain is the chief background. It is a long ridge achieving no more than 1,760 feet, but it is every inch a mountain, for, standing apart from the larger hills, it is a prominent landmark from the innermost parts of three counties and from large distances at sea. And it has the peculiarity of making a striking appearance from every angle. Its contours are slim and smooth. Yet those of its lesser neighbours are quite the reverse. Knobly carns of volcanic rock protrude from them, giving astonishing toothed outlines which in the distance make them appear like monster dinosaurs.

It was at a small bay in the Pencaer Peninsula that the last landing of an invading force took place within the realm of Britain. The event happened on an exceedingly fine day in February 1797. There are several accounts by contemporaries, each of them varying considerably. But the main facts are as follows. French ships of war appeared off Strumble Head and were observed to be standing in towards Fishguard Bay. There is, as I have said, a fine beach of sand at Goodwick, where it was obvious the ships would make for. However, before they could come to an anchor the alarm had been given to Fishguard, and two men, a volunteer artilleryman and a sergeant, had managed to run off to the little fort on the headland and had contrived to get two cannon into action, which was very creditable to them. This frightened the Frenchmen away from the good landing beach. They anchored, instead, off Careg Gwasted Point where they were unmolested. They spent the night in getting the stores for their base camp ashore.

Meanwhile the countryfolk took all their available horses and went inland towards the hills, while at Haverfordwest the trainbands were hastily mustered. General Campbell took charge of the situation and marched them off towards Fishguard. The landing-party consisted of about twelve hundred men under the command of an American-born citizen called Tate. The expedition was part of a joint enterprise to raise disaffection both in South Wales and Ireland. The Irish contingent, with Wolf Tone on board, was commanded by General Hoche. It made an equally successful landing, with equally unhappy results, at Killala Bay near Ballina.

It has been confirmed by recent history that an expedition of this nature will have a sporting chance in Ireland. But the intelligence department of the French had been sadly misled about Wales. They had come on the strength of a Baptist minister's assurances that Wales was Jacobin to the core, and that the Welsh peasants would rise to a war against their English landlords. It would then be an easy matter to secure Milford Haven. When this was done, Wolfe Tone's revolted Irish would pour across St. George's Channel, and there would be an all-Celtic march on London.

36 DIPHWYS FROM PANORAMA HILL, BARMOUTH,
NORTH WALES

37 BARMOUTH, NORTH WALES: the Mawddach Estuary and Viaduct

Tate's men behaved in a thoroughly systematic way to begin with. They brought all their stores, their baggage, their powder and shot ashore, and then went to look for horses and waggons to effect transportation towards Milford Haven. The horses were gone, but in the course of their ransacking they discovered more than one *cache* of spirits belonging to smugglers and their friends.

General Campbell had a force which more than twice outnumbered that of Tate. Nevertheless, to make doubly sure of alarming the foe he resorted to a picturesque stratagem which is not likely to be forgotten for many generations yet to come. He ordered that all the women (who then wore red cloaks and beaver hats) should equip themselves with broom-handles, walking-sticks, roasting-spits, or otherwise, and carrying these mock guns at the slope should parade in full view of the enemy at a distance just near enough to make it appear that they were soldiers in reserve. And the local boast is that by this ruse it was the women and not the men who saved England from Buonaparte. But, to my thinking, it is the smugglers who should wear the laurels, for the Frenchmen were much too drunk to see anything. They could not even reel forward when the word to advance was given, and General Tate handed them over in disgust. Not a shot was fired on either side during the land action (though at St. David's the Volunteers had been busy all night taking lead off the cathedral roof and melting it down for bullets). The only person who was the worse for the invasion was a Miss Anne Fenton, who, on hearing the first alarm, hid eight hundred guineas on the cliff top, before flying to the hills, and never succeeded in finding them again.

Near Strumble Head there are cliffs made of those curious hexagonal columns such as are found at Staffa and the Giant's Causeway. The ingenious natives of Pembroke have used some of these as gateposts. The coast which runs down to St. David's is full of lore about the past as well as visible relics of it. And one is inclined to believe that at a distant time—I am thinking as far back as the Stone Age—it may have been the most important part of Britain, if not the nucleus from which the Stone Age civilisation spread. For we are still confronted with that unsolved mystery of the inner circle of blue stones at Stonehenge which are now identified by geologists as having come from the mountain Prescelly which dominates this coast. And it may even have been a memory of some special sanctity believed to be inherent to this land which gave St. David's that pre-eminence on which it throve throughout the Middle Ages. No other bishop's city is so far removed from the centre of his diocese. When I see all the grand buildings which stand together at the end of this lonely and remote peninsula I cannot help thinking that they stand on foundations whose prestige goes back beyond the Christian era, is perhaps twice as old as that span.

Whitesand Bay is a beautiful strand just south of St. David's Head. Lying off the foot of the peninsula is the island of Ramsey, which until the Act of Disestablishment belonged to the Bishops of St. David's. Beyond it are a line of outlying rocks called the Bishop and Clerks. South Bishop Rock has a lighthouse known to sailors as the Bishops. This plural is an important one as it distinguishes this rock in the nautical mind from the simple Bishop, a lighthouse located in the Scilly Islands. All Liverpool ships coming from the Mediterranean or the South Atlantic base their calculations on either making the Bishop or the Bishops as their first landfall.

St. David's Head forms the entry on this shore to St. George's Channel with the South Bishop as its lamp-post, while, on the opposite shore, Carnsore Point is the entry and Tuscar Rock its lamp-post. There is a short gap of a few miles between this sea and the next. It is occupied for the most part by the splendid beach of St. Bride's Bay on the open Atlantic. The south shoulder of the bay ends in Wooltack Point off which lies the largest of a very interesting group of islands, all called after names which the Vikings gave them in the ninth or tenth century—Skomer, Skokholm (pronounced Skokeom) and Grassholm. The fishermen of Marlowes, which is the port on the mainland most concerned with the islands, still seem to bear about them the marks of a Scandinavian ancestry.

I visited these islands in 1915 before motors had made this coast accessible and before Mr. Lockley's writings had drawn the attention of the world to it. In those days the fishermen were extraordinarily simple and unsophisticated. Among other strange customs they retained one which caused them to raise their hats when passing a cave in Skomer Island. The reason given was that it resembled the female pubis; surely a curious survival of phallic veneration.

Grassholm is about nine miles away from the mainland,

38 THE CARDIGANSHIRE COAST AT CLARACH BAY, NEAR ABERYSTWYTH

39 THE OLD BRIDGE OVER THE TEIFI AT CARDIGAN

40 ABERAYRON: a small coasting Port on the Cardiganshire Coast

41 THE MOUNT: a Cardiganshire Headland

42 THE CARDIGANSHIRE COAST AT TRAETH SAITH (TRESAITH)

3 TENBY, PEMBROKESHIRE: the Sands and Castle Hill

and I have never reached it. Skomer and Skokholm are flat-topped with splendid cliffs and caves much frequented by seals. The rocky havens of Skomer, notably the Wick, are very striking. Fierce battles are waged here every spring between puffins and razor-bills fighting for the possession of the rabbit-holes as nests. Jack Sound, between Skomer and the mainland, is a fearsome place when the tide runs against the wind. Even worse is Wildgoose Race between the western extremities of both islands. Many small sailing-craft have been engulfed in it.

St. Anne's Head is the entrance to Milford Haven and also the northern pillar of the Severn Sea.

CHAPTER IV

THE SEVERN SEA

It is one of the little tragedies of geography that the name Bristol Channel has won its way into print on our maps instead of the older name for our principal western gulf, the Severn Sea. The change took place gradually in the first part of the eighteenth century when Bristol was in the height of its maritime power and prosperity. I have described the alteration as a tragedy because the new name does not supply the mind with an idea, whereas the older one gave the imagination not only a complete picture but, in addition, it gave it spare raw material to go on building with. And a name which gives food to the imagination is so much more useful to mankind than a name which is just a label. It seems fitting enough that our largest river should have a sea for its mouth instead of a mere estuary. That a sea should be called after a river confirms the whole of its coastline in something homely and landward. More important still, the Severn Sea is a romantic sounding name which might improve the aesthetic amenities of any coastline. In its own proper case, however, it does not idealise, it is simply an apt description by the subtle process of suggestion. There is no part of the coast of the Severn Sea which does not strike the full chord of romance in which picturesque historical events harmonise with rich settings of scenery.

The entrance to the Bristol Channel is laid down in Sailing Directions to be St. Anne's Head on the north and Hartland Point on the south. Also it has a lonely sea outpost, quite out of sight of land, being nineteen miles away from the nearest point (which is St. Anne's Head). This is a group of rocks which only appear at low water, called the Smalls. A lighthouse, which is very familiar to all southbound shipping out of Liverpool, stands here.

The first enterprise of putting a lighthouse on the Smalls must have appeared even more daunting to an engineer than that of the Eddystone Rock, for the site of operations is so much farther away from any base. It is not generally known that the first was built here, quite successfully, as early as 1775. The projector and proprietor was a Mr. John Phillips, and his engineer a Mr. H. Whitesides. The erection was mounted on nine pillars, three of which were of cast iron, the rest of wood. Whitesides, like the more famous erector of the first Eddystone Lighthouse, believed that it was his duty to prove the worth of his building by residing in it himself during the first winter that it had to stand the test of the stormy seas. This must have required some pluck, when Winstanley's melancholy fate[1] was common knowledge, especially when the Smalls is away from all human ken except that of the mariner.

It was the fate of Whitesides to be threatened not by a death from drowning but by one from lack of water. In fact, from various causes he and his companion ran short of water, food and even fuel for the light. The intrepid inventor then sat down and penned three exceedingly polite notes in the best manner of the period, stating the plight of the pair and begging for immediate assistance. The notes were addressed to John Phillips, the proprietor, who lived at Trelethin, a cove in St. David's Head. Each note was enclosed in a bottle, each bottle in a cask. The casks were hurled out of the lighthouse into the boiling seas on February 1, 1777, at a moment when the tidal current seemed to serve most favourably. One of these S O S bearers took a course right over St. George's Channel, down the south coast of Ireland and up the west coast, landing on the Arran Islands in Galway Bay in the beginning of April, when the Mayor of Galway took the best means at his disposal to get the letter forwarded to its destination in South Wales. The second barrel had a more fortunate voyage and landed in Porth Newgall in St. Bride's Bay. But the third cask went straight to the addressee. It washed up in Trelethin Cove right under John Phillips's house and was found within three days of its dispatch. So the heroes of the Smalls were saved.

St. Anne's Head not only marks the entrance to the Severn Sea but also that of Milford Haven. This remarkable stretch of water, like the Scottish firths and Killary Harbour in the west of Ireland, is a true fjord. That is, it is a sea inlet and not a river estuary, wherefore there is no bar across the mouth, which is the principal bugbear of estuarine harbours. It is a place that could accommodate several fleets at once, and its only drawback to development is that it lies right off the tracks both of trade and war. It has been tried for the Irish

[1] There is an account of this on p. 70.

trade and as a naval dockyard with only temporary successes. But the fishery industry flourishes. As a steam trawler base and a distributive fish market it rivals Grimsby and Fleetwood.

If Milford Haven had been mountainous like the Clyde or Killary Harbour it would probably have attracted more attention than any other physical feature in Wales. As it is, the shores of its wide main channel and many inlets are neglected by the rambler though they are full of local interest and, in places, of beauty also. A striking point about the scenery is that the rock formation alternates between limestone and Devonian sandstone, turn and turn about, giving unusual variety to cliff and shore contours. On one fine limestone peninsula stands the splendid ruin of Pembroke Castle, now, unfortunately, in process of being very much over-restored. From here went Stronghow's expedition in the twelfth century to conquer Ireland, backed by an official blessing from the Pope. And here, in the fifteenth century, landed Henry Tudor on his way to make a bid for the throne of England in the name of the House of Lancaster but fighting under the banner of the Welsh dragon.

Southward, outside the haven's mouth lies Freshwater Bay, a fine Atlantic beach that used to be deserted until the motorist discovered this remote countryside in very recent years. Then the coast turns inwards towards the east by a series of bold headlands of contorted limestone. The most notable of these are St. Govan's Head and Stackpole Head. By the former there is a famous fissure called the Huntsman's Leap where the cliff is riven by a narrow perpendicular cleft from the grass level to a sea gully below. But all this range of cliffs is delightfully freakish. Among their wonders are the towering Stack Rocks of Broadhaven and the palatially roomy caverns of Lydstep. Between Stackpole and Giltar Point there are a whole row of bays carved out of the grey wall as regularly as niches, among which are Freshwater East, Swanlake Bay, Manorbier Bay, Skrinkle Haven and Lydstep Haven. Each has a gleaming threshold of fine sand. Manorbier, in addition, has its castle, which is romantic in appearance and also by association. That interesting twelfth-century author, half Welsh, half Norman, Giraldus Cambrensis, came from here.

Giltar Point, off which lies Caldy Island and its small sister, St. Margaret's, opens the wide bight of Carmarthen Bay. The admiralty of Carmarthen Bay was one of the many appanages of the bishops of St. David's, and I think the title

44 LITTLE ST. GOVAN'S CHAPEL, in a Crevice of
St. Govan's Head, Pembrokeshire

45 THE MUMBLES LIGHTHOUSE AND BRACELET BAY, GLAMORGANSHIRE

46 THE ROCKY GLAMORGANSHIRE COAST NEAR PORT EYNON

still runs, though the bishop no longer gets any advantage from wrecks and shipping dues. The map shows how the head of the bay opens, letting in three winding prongs of water, like the claw of a Chinese dragon, into the heart of the land. In the Middle Ages an important and busy port was lodged within each of these prongs—Laugharne, Carmarthen and Kidwelly. An occasional steamer still finds its way up to Carmarthen, but I hardly think the other two ports have seen anything bigger than a fishing-boat for some time.

Caldy Island has fine cliffs of both limestone and sandstone. There are caves in the former, one of which bears the name of Paul Jones, the Scottish-born American privateer whose name is still honoured by commemoration at so many odd corners of the coast between the Solway and the Severn Sea. The floors of these caves have proved rich hunting-grounds for relics of the beasts and even the men of the Palaeolithic period. Caldy has seen all three phases of British monasticism. The Celtic saints had a culdee establishment here, there are the ruins of a mediaeval priory, and the Anglican Church built an abbey here in 1906. The monks, however, forsook their proper fold for that of Rome, whence they conveyed not only their souls but also the magnificent new buildings that had been raised to house them. Although the south of the island is high, it has a sheltered interior and grows excellent grass, on which many cattle from the mainland used to be sent to pasture. The method of conveying them to and fro is perhaps worth recording. You had an anchor cast out well ahead of the boat. Then you put a halter on your bullock. After which you heaved on the anchor. A barking dog assisted. Providing the anchor held properly and the dog barked sufficiently the bullock was soon afloat. The anchor was then hove up and the voyage begun. It is said that the bullock was not expected to swim and he seldom attempted to do so. He merely lay on his side and floated while the oarsmen did the rest. The journey is the best part of two miles.

Giltar Point not only opens Carmarthen Bay but also a different type of coast with long uninterrupted stretches of sandy beach. In fact Caldy Sound is the last bit of deep water that lies close to the land. From Caldy onwards the five-fathom line lies farther and farther away from the beach. East of Giltar Point the great cliffs cease giving place to the sandhills of Penally. Above this level the town of Tenby with all its buildings, mediaeval and modern, packed perforce on to a raised plateau of rock, rises with grand effect from the waves. This situation, chosen doubtless for defence against the sea raider, has in the latter days saved it also from the depredations of the land-raider. That is, it has largely been saved from straggling on one side at least. Tenby's cliff-top promenade is as ugly in itself as all others of its kind, but it affords one of the most magnificent marine prospects obtainable from any seaside resort, commanding as it does the whole sweep of Carmarthen Bay from rugged Caldy to the lofty outline of Gower.

Following round the Bay, by Saundersfoot and Amroth, the scenery is seen to have exchanged the grand austerity of the limestone cliff-face for something that is perhaps best described as *gentle*. Here, dell follows dell, each with its stream, some bracken-clad, some wooded, till you get to the end of Marros Sands when the limestone takes up the theme again—but not the same theme, for the cliffs of Pendine are not of the contorted but the slab type of architecture. That intervening space I have called gentle was an interlude of the coal-measures which yields only anthracite, and that in a limited degree. It is an unhappy paradox in scenery that the coal-measures invariably build a beautiful countryside and at the same time attract the most frightful forms of human spoliation and human habitation, accompanied everywhere by excesses in squalor—unless there is no coal in those coal-measures.

At Pendine the tide goes out for more than a mile, exposing a stretch of wonderfully compact sand six miles long. Racing motorists find this the nearest approach to the conditions of Daytona Beach that we have round our coasts. While the last word in modern mechanisms try their strength along this rink, and the fringe of the beach is crowded densely with spectators from the towns, there lies just behind the screen of sandhills a delightful arcadian solitude, seldom visited by the stranger, where patriarchal manners prevail. It is a little Holland with rich pastures held from the sea by dykes. Its population live in spacious farmsteads with red-tiled roofs and large white enclosure walls which look like monastic settlements. This is a bit of country that the sea has partly forsaken and partly been expelled from. From Ginst Point you may look back and see the old coastline standing up with its sea-worn cliffs and sea-battered caves two miles inland.

By Ginst Point is the entrance to that triple estuary I referred to a little way back as resembling the three-pronged claw of a Chinese dragon. If you sailed into this truly romantic entrance from the sea and made for the first harbour in the western cleft, which is the Taff estuary, the channel would take you with a circuitous flourish under a steep red sandstone headland 350 feet high standing sheer up on the starboard hand followed by a menacing escarpment of black rocks (on the same hand). Then a sudden vista would open on the far bank of an old seaport with half its houses on the shore and the other half raised up on a high rock platform. But the main feature is a castle whose courtyard and ground-floors are laid on the top of the rock but whose outer towers are founded on the seashore itself giving to the castle a singularly lofty grandeur worthy of an illustration from a fairy-tale.

This is Laugharne (spoken *Larne*). The castle which gives such an exceedingly feudal appearance to the ancient port does not belie the impression it makes. For Laugharne is still feudal. In fact its feudalism has saved it during more than one failure of industrialism. In the little town hall where they keep Sir Guy de Brian's original charter locked up, the corporation, with recorder, port-reve and bailiff hold a court (no longer competent to try criminals) at regular intervals, opening the proceedings with that familiar Norman tag, "O yes! O yes! O yes!" A quaint survival you may say; but there's more behind it. The land round Laugharne is cultivated on the mediaeval strip system, as in the days of good Sir Guy. It does not, however, belong to the lord of the manor but to the burgesses of Laugharne, about eighty in number. Each burgess holds fifteen acres in strip, free, for the duration of his life. Anyone who is born in Laugharne and has lived there for a stated period can establish a claim to burgage when a vacancy falls due. You can do a good deal with fifteen acres if you are clever. But you can't pasture on it as there are no hedges between the strips.

Had you sailed down the middle inlet instead of the western one you would have come into the Towy estuary. Here, too (on the port hand), within a mile of the entrance, you would find yourself almost within bowshot of another castle, quite different from Laugharne, but even more majestic in position and more picturesque as a ruin. This is Llanstephan Castle. The charming little village which has grown up under its protection lies nevertheless detached from it in its own narrow valley.

The blunt peninsula between Towy and Taf is a splendid countryside, such fields, such hedgerows, such farmsteads! Yet it is very little known because it lies on a track that is no longer beaten. Here is a remarkable instance of isolation caused by up-to-date communications. The modern route to West Wales goes through Carmarthen. The old route went through Kidwelly and came down to the Towy at Ferryside. Here there is an ancient ferry over to Llanstephan (the property of the owners of the castle). The journey is continued overland for three miles where a point on the Taf, opposite Laugharne, is reached. Here on a spit of green the road peters out and there is a building like a tiny church containing massive benches and a table. The wayfarer, having reached this building, tolls the bell, and when he succeeds in making himself heard on the opposite shore the ferryman puts off from Laugharne. No doubt, in old days, pilgrims to the shrine of St. David's used this route much. Within living memory it was the ordinary way for the Laugharne people to reach their nearest railway station, Ferryside. For even if the journey was only to Carmarthen it was cheaper to drive three miles, cross two broad ferries, and catch a train than to go by road to St. Clears and take the train from there. Llanstephan and the ferries throve on this posting business. There was nothing said or done either in Laugharne or Llanstephan or Ferryside that was not known in the other two. The strongest ties of gossip and of friendship bound the trio firmly together. Now the motor-bus operates between Carmarthen and each of the three, independently, and they have become, even within twenty years, as completely estranged from each other as if they were hundreds of miles apart. It is an instance of how modern transport can act as an isolating and uncivilising agent.

The Towy is still navigable as far as Carmarthen. It is a river which has a strong topographical personality from source to mouth. Its cradle is among the Elenith Hills which lie below Plynlimmon and are among the wildest and least known of all the hills of Wales. From the simple austere scenery of grass moors it descends into a country of dark gorges and gleaming cataracts—where Wales's most famous outlaw, Twm Shon Catti, had his cave. From here it issues into the upper level of a valley that in fertility must rank as chief in all the Principality, whence from salmon pool to salmon pool it goes brimming down to the tidal waters.

The third inlet (opening east of the Towy) receives the River Gwendraeth. It is the shortest of the dragon's claws and takes you in four miles to its head-water which is at Kidwelly. Here there is a third castle, larger and more magnificent than the other two, only it has not the advantage of their superb settings. It is, in fact, one of the grandest castles in all Britain, but its position is not striking. The size of Kidwelly church, formerly a Benedictine priory, testifies to a rich income from pilgrims. They probably came here by the shipload and made it a starting-point for the St. David's pilgrimage, using the two ferries mentioned above.

There is yet one more estuary in Carmarthen Bay. It is the most extensive in South Wales, though fathered only by a very small river, the Loughor. Here stands Llanelly, the home of the British tinplate industry, a very black outpost of the industrial area. On the south side of the estuary, however, there lies one of the freshest and most beautiful bits of country in the whole of the Severn Sea. This is the peninsula of Gower, which is famous not only for having a separate topography from the rest of Glamorgan but also for having a separate history and even a separate race. By the latter is meant an infusion of Flemish blood, which is true also of South Pembrokeshire.

The heights of Gower are under a thousand feet, but they are very striking in outline and look a great deal higher. At the end of the peninsula is Worm's Head, an island at high tide. In the distance, this dark rock mass with its tossed seaward peak is a great addition to the view of Gower. The name is said to be a Viking one, another form of Orme's Head, in fact. The south coast is fringed with limestone cliffs cut away deeply between headlands to form bays with generous sandy margins. The last of these Gower headlands is the Mumbles. Beyond that, the land takes a long sickle-like sweep, so clean that the edge of the coast looks as if it had been pared with a knife. This is Swansea Bay.

If you stand on Mumbles Head and look across Swansea Bay you confront a definite break in the scenery, the economic geography, and even the ethnology (speaking from a strictly modern point of view) of Wales. In short, we face the great industrial area. But we face it at such a safe distance that it gives us one of the loveliest views imaginable. Across the bay lies Margam Mountain, Mynedd Caerau, Cefn Mawr and many others, touching contours between one and two thousand feet, while stretching inland, to the north, rise the grander summits of the Vans and Fforest Fawr. Under this splendid background the long white training-walls of Swansea project into the sea. If you are lucky in getting one of those days of opalescent lights, not infrequent on this part of the coast, when the smoke haze from human industry joins issue with the mist about the mountains, Swansea itself will appear a sea city of Homeric glamour. The new civic centre, on the Southampton model, with immense pearly white halls and slender campanile (whatever you may think of it at close quarters) adds a telling *finale* to this striking paradoxical vista.

Swansea's prosperity during the last century and a half has been based even more than that of Liverpool on the Industrial Revolution. Her world fame has been built on the working up of metals of all kinds, made possible by the proximity of the great coalfield. But she would never have developed so successfully if she had not had deep roots in the past. The industrial boom found her well-equipped with a good harbour, a keen and intelligent corporation, and a wealthy and well-disposed lord of the manor. I hardly think that there is a seaport in the United Kingdom with a more instructive and satisfactory history except, of course, London, which remains incomparable because its advantages have always been unique.

There are no wide estuaries in Swansea Bay. The River Neath is navigable as far as the town of that name, famous for its iron-founding. Just at its mouth is Briton Ferry, with its works of iron and tinplate. On the eastern curve of the bay, under Margam Mountain, is another iron port, Port Talbot. It is built right among the sandhills at the mouth of a diverted river, a truly remarkable feat of harbour engineering. These same sands defeated the efforts of a whole people in an earlier generation merely to subsist on their own land. The once flourishing township of Kenfig lies foundered among a desert of dunes, a fragment of castle wall just marking the site. It is the gravestone of a community.

The coastline now bears into the sea with a wide convex front. Inland of it lies a rolling platform of fat agricultural land some ten miles deep, stretching from the feet of the mountains, famous throughout history under the somewhat misleading name of the Vale of Glamorgan. It was the first valuable bit of Wales the Normans succeeded in annexing Fitzhammon and his twelve knights sailed out of the Bristol Avon in 1093 and simply planted themselves along this crescent of coast. Cardiff, the leader's prize, was the headquarters of the colony. Castles that have been several times rebuilt still mark the original sites chosen by the twelve. The best preserved of these, though it retains no trace of Norman work, is St. Donat's. It stands overlooking the sea just where the coast turns to face due south. Which of Fitzhammon's followers first staked the claim and put up his mound and wooden castle is not known, but the place has been inhabited continuously, and is now in the ownership of an American newspaper magnate. A strange story is told of the last member of its longest line of owners. By 1738 the family of Stradling had been at St. Donat's for four hundred years. The long gallery was hung with their portraits, five or six generations deep. But in the year mentioned, the last male heir of the family was killed in a duel on the Continent. They contrived to bring the body home. It lay in state in the long gallery, where the family portraits were made to share with it the light of corpse candles throughout the eerie watches of a stormy night. By some misadventure the tapestry took fire, and round that dead body which bore the last of the name the images of the whole ancestral host were consumed.

This coast is free from industrialism until you reach Barry. It is not so grand and striking as that of Pembroke and Carmarthen but it is pleasantly varied. It has fine cliffs in places made of the blue-grey lias rock. The most striking of them is Penarth Head where veins of alabaster appear. I fancy Penarth Head is a sailor's name, for it is shown on the maps as Lavercock Point. Barry, which is just on the seaward side of the Point, has a wide inlet with an island at its mouth. It is an unusual island for such a situation, for it faces the sea with a deep and wide crescent which holds a beach of beautiful sand. Now, while all the creek behind the island is built up with an immense dockland and sternly industrialised, the beach is given up to the pursuits of pleasure. In the old days the island with its conveniently concealed harbour formed a stronghold for the dreaded pirates who preyed on the Bristol shipping. When the seas were made safe from these pests the smugglers took over their old quarters at Barry. Tales of one Knight, the last of the great "free traders" who was finally chased away to Lundy, are still told in the neighbourhood.

Penarth itself lies just within the sweep beyond the headland, close by the entrance to Cardiff Harbour. Cardiff has been a place of military importance since the time of the Romans and has always been a port of some note. It is mentioned as supplying ships for Edward the Third's great expedition to France. But its elevation to the rank of a first-rate port and subsequent civic claim to being called the capital of Wales dates from the late eighteenth century when the great coalfield which lies behind it was put into contact with the sea by a canal, and in the early days of the next century a dock was built for the necessary operation of transhipment by the enterprising second Marquis of Bute.

The success of the Bute Dock was sudden and sweeping and it was not long before it appeared to coal-owners in the light of a monopoly whose burden was irksome. It was owing to this challenge that Penarth and Barry came into being. One can imagine the temperature of the rivalries when these two near neighbours opened their dock-gates and coal-shoots to the world of shipping. But now they are all under the ownership of the Great Western Railway and the economist and not the local patriot is the arbiter of their fates.

Penarth Head is a definite boundary between the wide waters of the Severn Sea and the narrow ones. Opposite to it the Somerset coast takes a right-angled sweep round Bridgwater Bay and moves due north towards Wales. In the background the steep limestone ridge of the Mendip Hills which stands above the flats of Sedgemoor aligns its axis on Penarth Head as if determined to join it. But the range breaks down in gaps as it approaches the shore, throwing up three hills, the last of which, Brean Down, 305 feet high, stands out into the sea as a peninsula which is almost cut off from the land. It points straight to Penarth Head, and, in the interval between them, two islands rise from the sea bearing the Norse names of Flatholm and Steepholm, which are descriptive of their actual appearance. On the former grows our rarest ranunculus, the wild peony.

Into this inner haven of the Severn Sea four great rivers fall, the Severn, the Wye, the Usk, and the Avon. The Severn's port is Gloucester, far up round snaky bends. Ships reached it within living memory by being man-towed up the final reaches. They were mostly small coasters of a local type called "skows." Now there is a ship canal, entered at Sharpness, which takes steamers into the Gloucester docks.

47 LAUNCHING A MOTOR BARGE AT BRISTOL DOCKS

48 BULL POINT, with its Lighthouse, near Morthoe, North Devon

The Wye, born within a few yards of the Severn on the slopes of Plynlimmon, unites with its big sister just within the bounds of her estuary, having displayed through its course all the characteristics of an opposite personality. The Severn, languid and muddy, wanders with only one exception through uneventful horizons. The Wye, rapid and clear, flows, from beginning to end, through scenery of the most dramatic character crowning the achievements of its career as it passes out of its wooded gorge to meet salt water by the ruins of Tintern Abbey and under the castled crags of Chepstow. The Usk, a stream hardly less famous for its beauty than the Wye, meets the tidal waters by Carleon, the town of the Second Legion of Rome, and the second city of Britain in King Arthur's day. A little farther on it serves the shipping at Newport, whose coaling berths are developed to the last pitch of modern efficiency.

The Avon cleaves the Cotswold Hills, is canalised at Bath, and joins the Frome just outside Bristol to form the "floating harbour" of that great port. Afterwards it passes through the steep, spectacular gorge of Clifton, reaching the Severn Sea at Avonmouth. The story of Bristol is one of exceptional interest, for it won commercial freedom from its feudal lords as early as the thirteenth century. Its fellowship of Merchant Adventurers was one of the most powerful trading concerns which England ever produced before the days of private firms. Bristol's first trade was probably with Ireland. It soon spread to the Continent and the Mediterranean. The wool industry of the Cotswold Hills came to the looms and ships of Bristol. While the discovery of the West Indies was made by Christopher Columbus in 1492, it was John Cabot who made the first discovery of the mainland of North America five years later, his venture sailing out of the port of the Avon. For long Bristol was supreme in the trade of the Guinea Coast and the West Indies. But in the eighteenth century the engineer, Telford, writes, "Liverpool has taken firm root in the country by means of the canals: it is young, vigorous, and well situated. Bristol is sinking in commercial importance: its merchants are rich and indolent, and in their projects they are always too late. Besides, the place is badly situated." A little later, in 1835, Bristol sent the first steamship across the Atlantic—the *Great Western*. But whereas she was content to attempt the new trade under steam with one ship, the Cunard Company of Liverpool followed suit with four steamers and secured the mail contract and made Liverpool the principal passenger port for the United States. As to Telford's objection to the situation of Bristol, it was indeed a drawback to have to tow ships up the breathless gorge of the Avon by crews in rowing-boats—even when the tide assisted the operation. Steamers did not suffer the same handicap. But the Avon Gorge was Bristol's barbican in the early days and saved her from those raids by enemies and pirates which set back the growth of so many of our more exposed ports. After the river had been converted into a floating harbour by penning it in the heart of the city within dock-gates all went well for a while. But with the increasing draught of ships the Avon bar gave trouble in spite of all dredging. Now all these difficulties have been solved by the establishment of an out-port at Avonmouth on the open water.

The Severn estuary begins officially at the roadstead between Bristol and Portishead, called King's Road. If you would know how strong and individual is the power of poetry of this estuary as compared with that of others you should stand on Beachly promontory, just above the exit of the Wye, at the point where the great earthen dyke raised by Offa, King of Mercia, ends its hundred miles of transit from the Cheshire Dee. You should stand here above the ancient fish-traps and watch the water flowing in the channels and gullies among the sandbanks, encircling the great ribbed expanses, rising over them, gathering itself together to rush with its bore wave right up to Tewkesbury. And as you watch, listen! There is no localising the sounds of murmuring, boiling, rustling, twittering. But as the tide brims over the flats, so these sights and sounds brim over the levels of conventional thought and understanding bringing an impulse from the unfathomed.

The coast round Bridgwater Bay is also in the minor key but in a different scale. Over the green pastures, the reedy dykes and the red osier beds of Sedgemoor you may look one way and see the Tor of Glastonbury on the ancient Isle of Avalon rising up startlingly pointed and luminously blue. You may turn to another direction and look over ungaugeable distances of level lushness to where a solitary figure in silhouette breaks the low horizon—the ruin of a church filling the top of a hill as evenly shaped as an inverted bowl. It is the Isle of Athelney where Alfred burnt his cakes

Immense sands stretch across the opening of Bridgwater

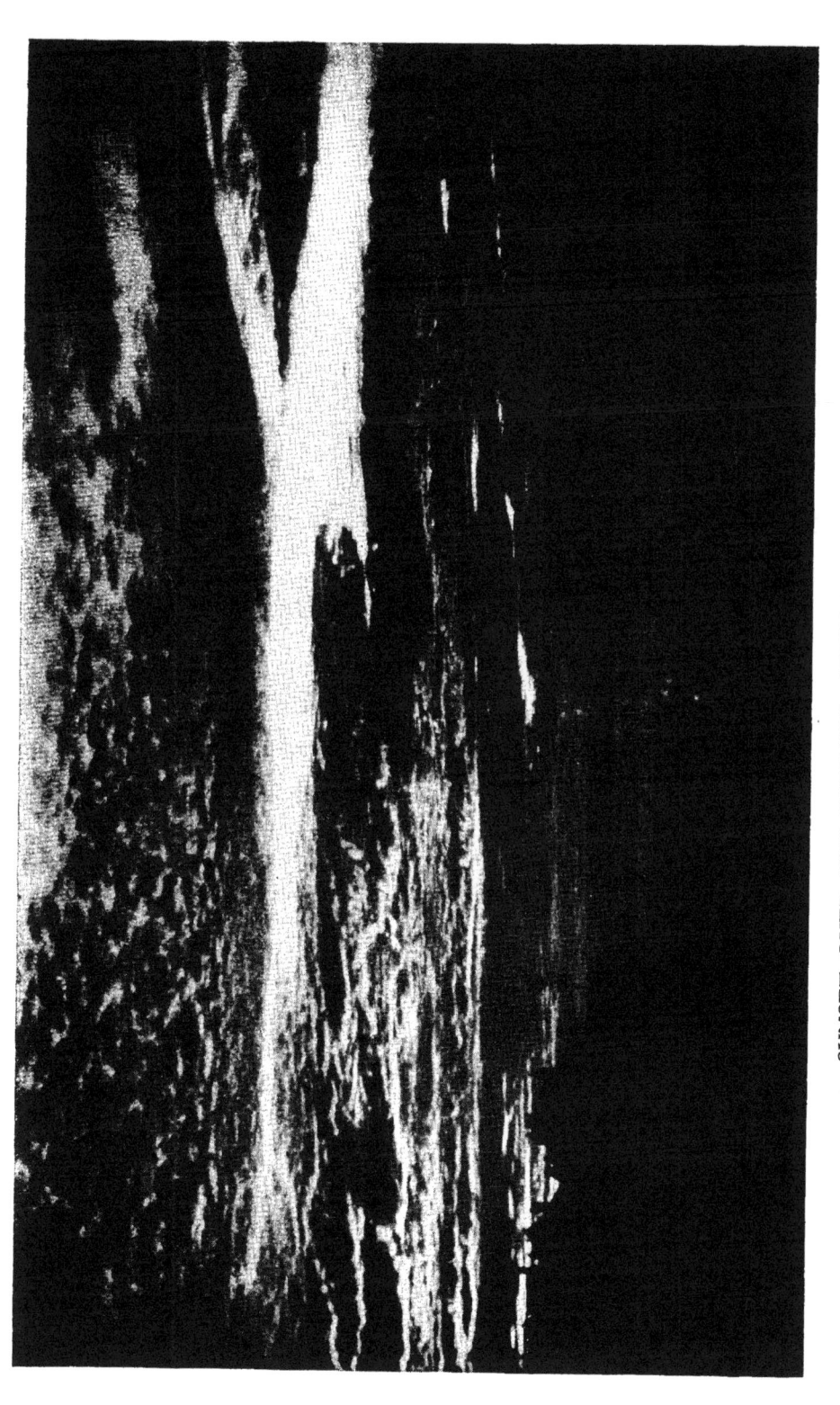

49 SUNSET OVER THE ESTUARY, APPLEDORE, NORTH DEVON

50 THE LITTLE HARBOUR OF LYNMOUTH, NORTH DEVON

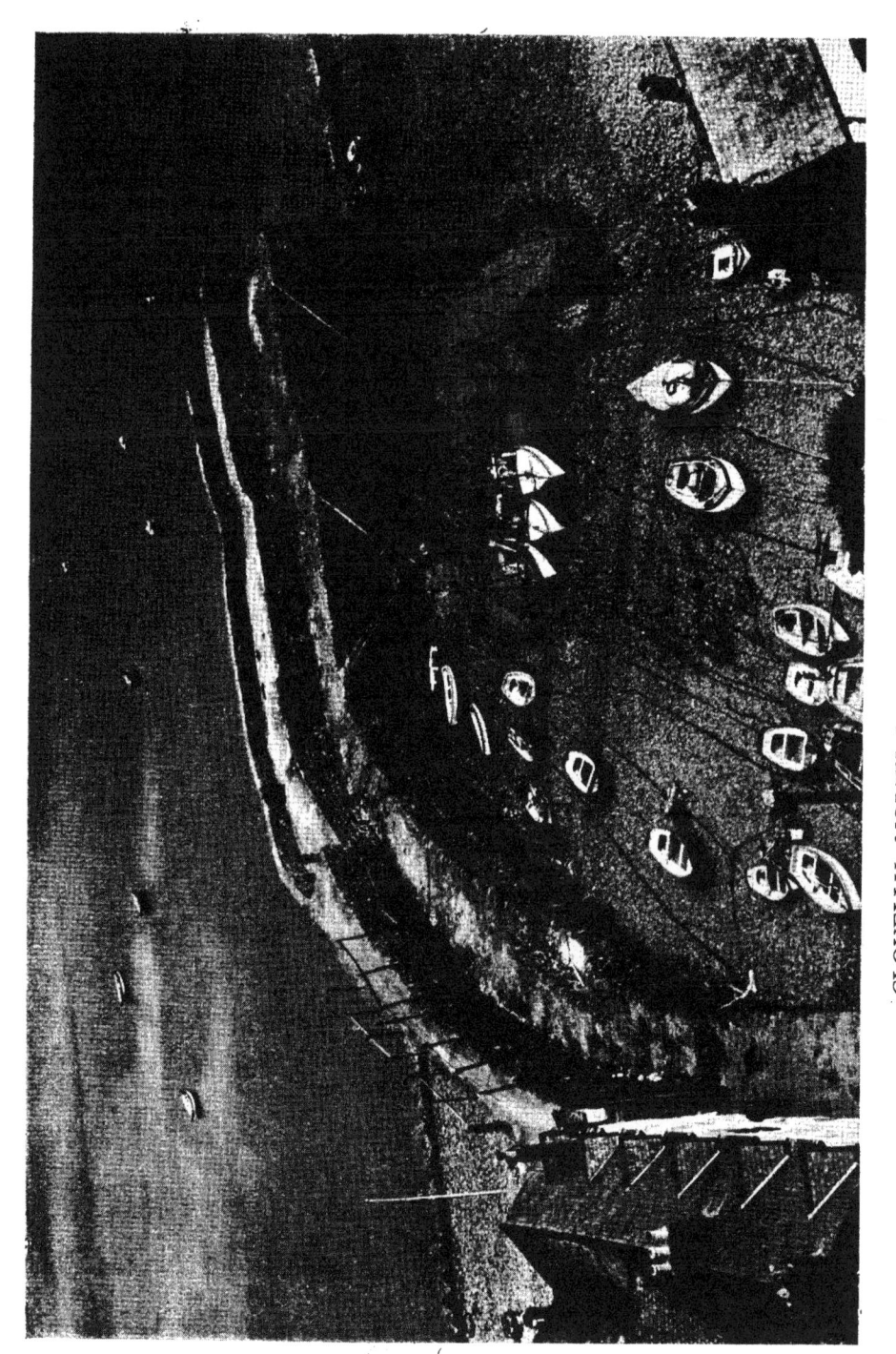

51 CLOVELLY, NORTH DEVON: the Harbour at Low Tide

52 A GLIMPSE OF THE ROCKY DEVON COAST
AT ILFRACOMBE

Bay in the midst of which the River Parret, famous in the annals of Sedgemoor, struggles to the sea. Shipping can still get up it to Bridgwater on the top of the tide. On the west side of the bay the coast rises in flowing outlines, ridge upon ridge, into the closely packed contours of the Quantock Hills. Though small in scale, this is one of the prettiest of our ranges, red sandstone cliffs rising to bracken-clad slopes, capped by rolling tops of heather. The ridges lie parallel to the coast, but the crests rise and fall and they are indented with frequent dells bearing woods and streams. The charm of the Quantock villages which just peep out from the cover of the hills to the sea is incomparable. Round the recess of Blue Anchor Bay stand Watchett, and Cleeve with its beautiful ruined Cistercian abbey made of the warm-tinted local stone, Dunster with its still inhabited castle, and the town of Minehead. Here begins the long upward slope of North Hill which rises to more than a thousand feet in Selworthy Beacon standing over Porlock Bay.

Thereafter follow the heights of Exmoor. Just over the Devonshire border, Foreland Point stands out into the sea surmounted by an imposing crag. It forms the eastern side of Lynmouth Bay.

I fancy it would be futile to attempt to add to the praises of this astonishingly beautiful region which stretches between Lynton and Ilfracombe. Here lie Woody, Elwill, and Coombe Martin Bays. By the last named, the Great and Little Hangman hills rise above bold pedestals of red cliff separated from Huddlestone Down by a deep gorge carrying its stream to the beach.

At Bull Point with its sturdy lighthouse the coast begins to turn south. A little farther on, at Morte Point, the scenery changes to a sterner order as the coast swings to face the Western Ocean with a typical Atlantic beach at Woolacombe Bay. Baggy Point is a favourite haunt of the buzzard and peregrine. In Bideford Bay the cliffs and high ground retreat inland to form the V, in the nick of which lies Barnstaple Town, while Saunton Sands and the great warren of Braunton Burrows stretch across the mouth of the Taw estuary. Just within the mouth of the estuary stands Appledore where the Torridge, Bideford's river, comes down to join the Taw. Bideford, the heart of Kingsley's *Westward Ho!* has still a lively quay. From here a fishing-boat takes the mail to Lundy.

Lundy Island is a tableland of granite which rises precipitously out of the sea. It is steep in every part except the south, where Providence has given it a small landing beach of sparkling mica-strewn sand. It was held (or was, when I was last on it) direct from the King and was not amenable to the acts of the British Parliament. Here you could safely defy the education officer and the income-tax collector. You could land a cargo of rum or tobacco there which had come direct from their ports of origin, under the nose of the coastguard (I am speaking of 1914) and he couldn't make you pay duty for them. But I fancy these old privileges have been somewhat curtailed since the King of Lundy began minting a coinage of his own.

It is a wonderful place for pure sea breezes. The lofty table-top wears a green cloth of fine pasture. But careless kine are not infrequently blown off the edge into the sea four hundred feet below. One of the sights of the island is to watch the diving of the gannets. They hover over their finny prey on the level of the cliff-top (quite close to you, if you sit with your legs dangling over the edge), then fold their wings and drop headlong into the abyss, diminishing in size as they fall. Through the clear water you can still follow them as they make their capture.

Famous picturesque Clovelly, with its steep streets, its much photographed old ladies and donkeys, is on the west side of Bideford Bay which rises in formidable dark cliffs made of millstone-grit. It culminates in the lofty rock of Hartland Point whose stern features are cheered by the dazzling white buildings of a Trinity lighthouse, lodged half-way down the face. Here lies the southern official limit of the Bristol Channel.

53 LOW TIDE AND CRAWLING SEAS AT SAUNTON SANDS, NORTH DEVON

54 WIDEMOUTH BAY, CORNWALL: an Atlantic Sunset

CHAPTER V

THE ENGLISH CHANNEL

The "Cornish Sea" is a Tennysonian, not a cartographical entity, which is as well for the romantic and holiday mood, as it tends to preserve Arthurian horizons from the violence of antiquarians. If we place it between Hartland Point and Land's End it does not technically fall within the bounds of any of our five narrow seas but belongs to the seaboard proper of the Atlantic Ocean. Now, the coasts of the Cornish Sea and English Channel have received far more attention from authors, publicists and public than any other part of our coastline. So I hope I shall be excused if I do not treat these shores in such detail as the others. It is a far cry from the Solway to the Tweed via Land's End, and one must take short cuts where one can!

Cornwall is a Celtic country which until the year 577 was one with Wales, enlarged in those days by a now vanished province called Strathclyde which reached up along the eastern shore of the Irish Sea, the North Channel and the southern shore of the Firth of Clyde. Strathclyde has been wholly absorbed by the English and the Scotch, but Cornwall has maintained its identity with the Welsh nation in nearly everything except the language, and this has only perished within the memory of the living generation. Dolly Pentreath was buried with honours before my time. She may have been the last fluent speaker of Cornish. But an old postman at Zennor, who was alive in my boyhood had a good enough grasp of the language (by memory, and not by book) to give that extension which I have claimed for Cornish as a living language.

But the types of people in Cornwall differ greatly from those in Wales. Their legends and folklore (alas! how distorted and exaggerated by publicists and cranks) are kindred, but deal with bolder characters and are altogether of a more colourful fantasy. They have always been much less fearful of the sea than the Welsh and their stories of wrecking and smuggling are more daring, vivid and picturesque. For, whereas, the Welsh became a nation of herdsman and warriors, the Cornish were ever occupied with mining and seafaring.

It is this fact which gives the coast of Cornwall an interest which must always be added to that of the scenery. It is a county whose coastline and coast-dwellers are its glory.

The North coast of Cornwall has its grandest headland at Tintagel which, crowned by its ruined castle, rises nearly severed from the land. Its grandeur is augmented by the moorland tracts which lie behind it reaching up to the bald summit of Brown Willie (1,364 feet), Cornwall's highest hill. The only large inlet on this coast is at Padstow where the River Camel takes the sea easily in a wide curving estuary. Boscastle is a rock harbour at the mouth of a delightful little trout stream called the Valency which comes down from the moors through a deep valley beset with woods and brake and orchards hoar with mistletoe and old-man lichen. Bude and Newquay have wide frontages of hard sand. Port Isaac is a cliff cove. St. Ives, deeply embayed from the forceful, western impulses of the Atlantic, has a tolerably safe harbour. From the sea, at the eastern lip of the bay, rise the two great tilted rocks of Godrevy surmounted by a sturdy white lighthouse. This shore has a range of mountainous sandhills where ragwort and bugloss make the bravest show imaginable among the shimmering marram grass.

In contrast to the heel of England, which is shod with friable chalk, her toe is shod with granite, the most sea-resisting of all rock. This peninsula between the two bays which shelter St. Ives on the north and Penzance on the south, is one of the five great granite bosses which rise up through the hard, contorted crust of Cornwall, and the only one exposed to the carving tool of the waves and the hammering of Atlantic storms—a small area, though you might call it the most *intense* part of Cornwall. Here reside the most striking features of its coast and moorland scenery. Here, also, is a vast concentration of megalithic monuments, the dolmens of Chun Mulfra, Lanyon and Zennor, the Men-an-tol and Maen Scryfa, the very remarkable circle of Boscawen-Oon, hut-circles, logan rocks and other marvels which have a true and alleged association with that powerful, superstitious race of the Stone Age. Not the least remarkable of the physical features of those high moors are the *carns*, wind-sculptured turrets of granite (such as are called *tors* in Devonshire and East Cornwall). Notable among them are Carn Galva and the Hooting Carn which cut the weirdest figures in a sea fog. From these uplands you may look all round on to the infinite

55 THE RUGGED GRANITE OF THE CORNISH COAST NEAR LAND'S END

56 ST. MICHAEL'S MOUNT FROM THE SHORE BELOW MARAZION, CORNWALL

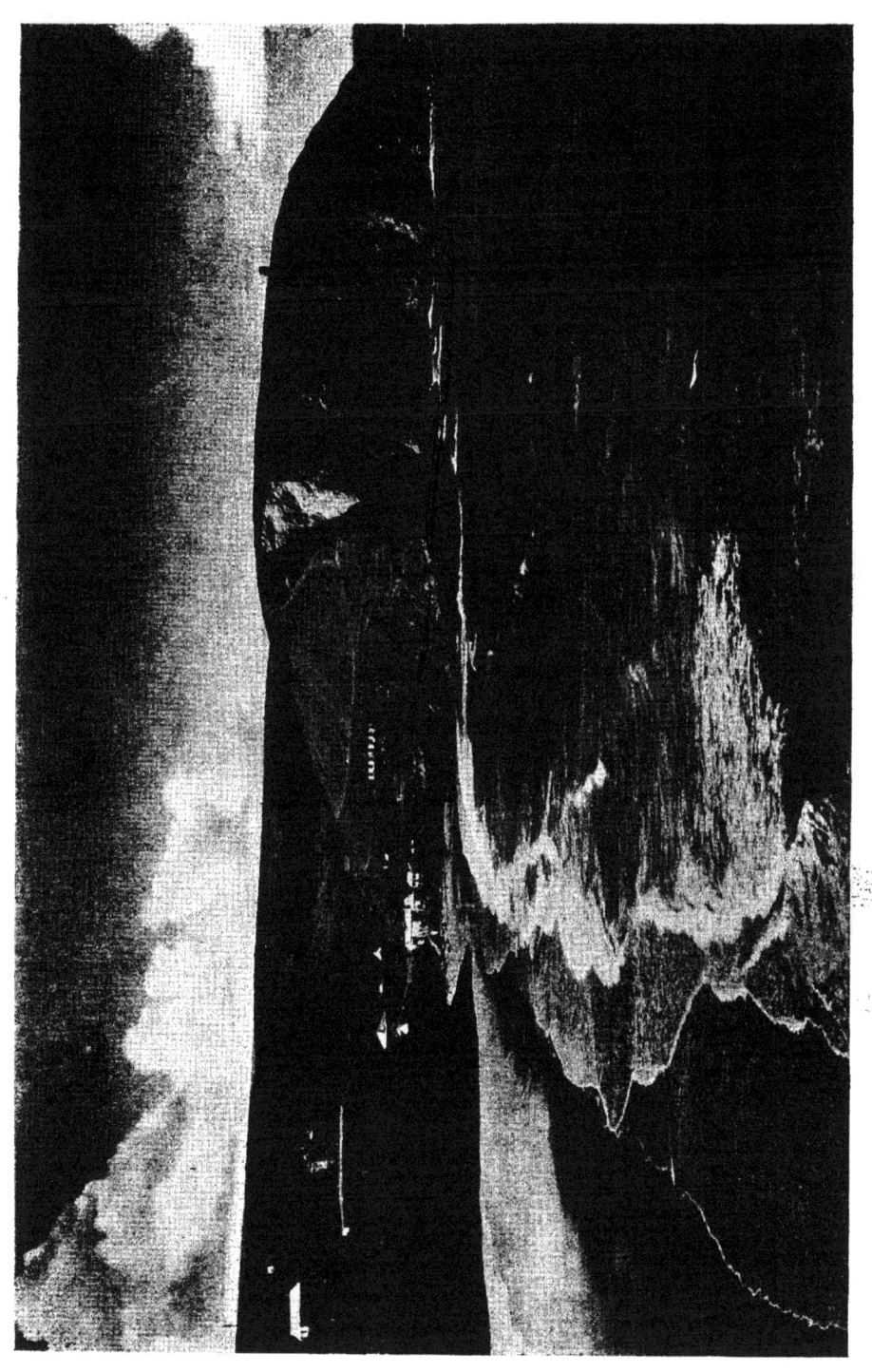

5 AMY'S COVE AND THE WESTERN HILLS FROM THE BEACH AT PORTREATH, NORTH CORNWALL

58 TINTAGEL CASTLE: a ruined Stronghold of the Cornish Coast

59 PORTHLOE: a Village-Port of South-East Cornwall

60 THE SANDS OF PERRANPORTH, NORTH CORNWALL

61 THE CORNISH COAST NEAR CAMBORNE

62 HARBOUR SCENE AT ST. IVES, CORNWALL

plain of the Atlantic, whose blue is modulated in fine weather by long stripes of deeper colour, where the rollers travel shoreward, and by skeins of white where the tide-rip churns. To the north and west sea- and sky-scape blend, but the southern horizon is embossed with the cluster of Scilly which looks in its aerial perspective as fabulous as the figures of romance it has evoked about the lost land of Lyonesse, whose remnants still threaten the mariner by those dots on the blue —Wolf Rock, Runnelstone and Seven Stones. Everywhere on this inspiring upland the feast of the eye is helped to a higher plane of enjoyment by the challenging scent of the moorland plants blent with briny wafts from the open main.

Among the many striking features of this peninsula, at sea-level, is the leonine rock cape of Gurnard's Head, the toppling cone of Cape Cornwall, the fantastically rugged group of islets called Longships which, from Land's End lie in the eye of the sunset, and the magnificent granite cliff of Tol Pedn Penwith. Some of the choicest coves of Cornwall, are nooked out of that famous fringe of cliff, and there is a bay here whose like I have only seen twice—in Ireland and Brittany. It bears the true but commonplace name of Whitesand. It is composed of white quartz grains and sparkling flakes of mica. When distant storms begin to brew this bay is the first to give warning by a thunderous swell which echoes far inland. But its beauty lies in the way the clear blue-green water from its wave-wash lies a bare instant on the sparkling whiteness of the sand and then clean vanishes into its porous mass.

In the crook of Mount's Bay, Penzance and Newlyn shelter. The steamer to the Scilly Isles goes from here, and here the herring fleet from Lowestoft and Yarmouth come and forgather with the native fishermen. Raising itself from the inner waters of the bay, one of the chief ornaments of our coastline, stands St. Michael's Mount topped with the towers and roofs of its old castle and monastery, now inhabited by the lords of St. Leven. It is lesser sister, both in feature and blood relationship, to Mont St. Michael across the way on the coast of Normandy.

Mount's bay is closed by Lizard Head, famous for its lighthouse, its cliffs of serpentine and its smuggling past which centres in the King of Prussia Cove.

The most notable feature of the wide bay which stretches from Lizard Head to Start Point in Devonshire is the number of its inlets of a peculiar kind, something between fjords and estuaries. That is, they are river mouths without having the usual disadvantage of dangerous bars. The largest are the great natural harbours of Falmouth and Plymouth. Fowey is situated on a third, Saltcombe on a fourth (this, oddly enough, has a bar, but no river). Perhaps the most remarkable is that at Dartmouth, which is just outside the area we have mentioned. But the most *sporting*—if one can apply such a jaunty term to an element of physical geography — is the one known as the Helford River. This cuts into the moor-clad peninsula of the Lizard from its eastern side. The Helford River has the general characteristic which marks all the others, namely, a deep channel leading from the sea snugly into the land with a ramification of tributary creeks branching off. It looks as if it had been intended by Nature to afford a convenient exit and entrance for the trade of the country. But the tiny village of Helford is the only thing resembling a human settlement which stands upon its banks. Its ample amenities would have been lost upon the seafarer had it not been for the yachtsman, for whom it is an ideal destination.

Falmouth Harbour is entered between Pendennis Point and St. Mawes Head, each of which bears one of those coast-defence castles built by Henry the Eighth—the last royal castles to be built in England. The town is situated finely just within the harbour mouth. It stands on rising ground above the quays and docks. In the anchorage the old tea-clipper, *Cutty Sark*, remasted and re-rigged, lies to moorings in her last berth. One of the four long arms of the great inlet goes up to Truro, the capital of the county. Above the shores rise wooded and agricultural slopes high enough to give that beauty, if not grandeur which is lacking at Milford Haven. At King Harry's Passage the water runs a dark jade-green between steep shores thickly embowered with overhanging trees.

Falmouth is nearest to the principal tin-mining area of Cornwall which is situated all round the great boss of granite whose summit is the bald hill of Carn Brae, visible from far off at sea on both sides of the peninsula. The old sea port of Fowey, made famous by the writings of Sir Arthur Quiller-Couch, is the chief port of the china clay industry. Just east of Fowey lies Polperro, one of the most picturesque rock-harbours in the West Country.

From seaward, by the Eddystone Rock, Dartmoor appears

63 CROSSING THE BAR AT BUDE, CORNWALL

64 POLPERRO: a Cornish Fishing Port

65 PLYMOUTH : Moorings for Fishing Craft at the Barbican

like a blue mountain rising above the entrance to Plymouth Sound. If your definition of a mountain is a hill of not less than two thousand feet high, Dartmoor qualifies with thirty-nine feet to the good (at Willhayes Tor). It is, in fact, the only mountain on the South Coast. To the many rivers which rise in the swamps of its rugged uplands Devonshire owes much of the beauty of her scenery.

I think it cannot be disputed that Plymouth is our finest and also our most beautiful harbour—unless you put the Lower Clyde into comparison. It is really compact of two natural havens and one semi-artificial one. By the latter I mean the Sound, which is the principal anchorage for the large liners which call at the port. This is sheltered by a long isolated breakwater which lies athwart the entrance and has a wide passage at either end between itself and the land. From the Sound, two arms of the sea join river mouths and run into the land, forming Catwater (the Plym estuary) to the east, and Hamoaze (the Tamar estuary) to the west. Three towns stand on the promontory, facing seaward, where these havens converge, Plymouth, Stonehouse and Devonport. The three have for long been united with links of pavement, brick and mortar. Lately, they have become more humanly bonded, for the Three Towns have united to form one city.

At the western entrance of Plymouth Sound the Cornish coast sweeps round from Looe Bay to the conical promontory of Rame Head with jagged fissures of cliff at its foot and a ruined chapel on its summit. On, beyond the breakwater, in the still, clear deeps of the outer harbour, wooded Mount Edgecumbe rises above Drake's Island, and Plymouth with its Hoe, its terraces, and its citadel lies right ahead. The entrance on the eastern side of the breakwater is equally fine. Here you pass under the cliffs of Stoke Point at the end of Bigbury Bay and cross the mouth of the Yealm River with its pyramid island sentinel, the Mewstone.

Plymouth suffered at least once the fate which seems at some time or another to have overtaken all the Channel ports, namely a sack and burning by the French. It was the chief naval base in Elizabeth's time, and from here Drake's contemptibles made their sally on the Spanish Armada. But there is surprisingly little left of Elizabethan Plymouth, though those who know it well say that the spirit of those days can still be *sensed* in the neighbourhood of the Barbican.

It was a port of call for the *Mayflower*—the last one. And it is to-day chiefly interesting as a port of call and not as a port of destination.

The lighthouse at the Eddystone is the fourth to stand on that dangerous submarine pinnacle, which barely puts its nose above the water, twelve miles from the nearest land. The first was built in 1696 by Winstanley, a private individual who must have had both outstanding genius and courage. Besides the ordinary difficulties of construction on a site which was awash at high tide (as you may read of in Smeaton's journal of later date) Winstanley was captured by a French privateer and carried overseas to a dungeon as a valued prisoner of war. An application by our Admiralty, however, secured him immediate release. He was, as a matter of fact, exchanged, but Louis the Fourteenth is said to have intervened personally and to have speeded his parting guest with the remark "I am at war with England, not with humanity," an epigram that is always taken to be magnanimous, but nevertheless bears a painful interpretation as a *double entendre*.

Winstanley's lighthouse was a wooden tower which rose a hundred feet from the rock and had many ornate and remarkable amenities such as a bow-window for the light-keeper to fish from. In 1703 the enthusiast paid a visit to his lighthouse to carry out some small repairs. While he was there a storm of unusual violence struck the coast. The folk in Plymouth were deafened by the noise and rocked in their their beds. They rushed to the Hoe the next morning at break of day to look seaward. Peering through the haze of spindrift that teemed off the hoary main, they perceived that the offing was empty. The great tower on the Eddystone was completely gone.

The next to try was also a private individual, John Rudyard, a silk mercer. He also built a wooden tower but anchored it to the rock more scientifically than Winstanley. It lasted for half a century and then took fire. A rescue party was on the spot before the next tide rose. They found the light-keepers huddled in terror on the rock itself. One of them, an old man of ninety-four complained that while he had looked up at the blaze, open-mouthed, he had swallowed fire. This was thought an amazingly funny story and went the round of the town. But the old gentleman succumbed to the shock and the surgeon who held the *post mortem* recovered from his stomach a piece of solid lead weighing seven ounces and five

66 LADRAM BAY, a little Devon Cove near Sidmouth

67 THE BROAD ESTUARY AT SALCOMBE, SOUTH DEVON

68 A LOBSTER-FISHING VILLAGE: the Shingle Street at Beesands, South Devon

69 BRIXHAM, SOUTH DEVON, FROM THE HARBOUR

drachms. Smeaton was the third builder. His tower has been dismantled, taken ashore and rebuilt on the Hoe.

As there are no inclusive names for the indented cape where the coast of Devonshire turns from a south-going to a north-going direction or for the wide bay which lies between Start Point and Portland Bill, I would suggest that they might aptly be called respectively the Promontory of Promontories and the Bay of Bays. The former is made up of Bolt Tail and Bolt Head, Prawl Point and Start Point. The main headings of the latter are Start Bay, Tor Bay and Lyme Bay. These may all be sub-edited into bays and coves innumerable. Nothing is harder than to make comparisons in our coastline. But Turner said of Plymouth that he had never seen so many natural beauties in such a limited space of country as he saw there, and we may perhaps say of the coastline just indicated that, taken all round, for what is to be admired in a mingling of land and sea, it is our premier stretch of seaboard. These headlands have been the leading landfalls and departures for all seamen who have come and gone from the North Sea and the upper Channel ports towards west and south. Holyhead, the Fastnet, the Downs, the Lizard, the Bishop and the Start are names known to all deep-sea sailors of all nations and languages throughout the world.

Eastward of Start Point the forces of the Atlantic abate, and the English Channel begins to assert its own individuality as a sea. Let us briefly anatomise it. It begins officially at the south-west point of Cornwall and the westermost tip of Brittany. The waters between these points are called the *chops* of the Channel, meaning lips (as in Bath chop) and not choppiness, in the sense of rough water. The imagery of the sailor nearly always works out as nicely as that of the true poet. The chops of the Channel snarl and show their teeth—on our side of the jaw they are the Longships and the Scillies, on the French shore the principal fang is the Isle d'Oessent, always known in sea parlance as Ushant. Within the jaw, Devon and Cornwall are confronted by the wide Bay of St. Malo. The whole of this area, where clear blue water fresh from the springs of the Western Ocean rolls gallantly in on splendid beaches, is subject to a large rise and fall of tide. But the sea becomes a confined one between Cap de la Hague and Portland Bill, and the tide from this arm of the Atlantic has to reckon with the undulation which comes down our east coast to meet it in the Straits of Dover. This causes two peculiar conditions, one is the curious halting flood which mounts the Solent in two impulses and then keeps high water stationary at Southampton for two hours, the other is the eddy in the Straits of Dover which, judged aright, will enable a swimmer of endurance to cross the Channel without having to make part of the distance by dint of muscle. Except in the neighbourhood of Folkestone and Dover, the sea has not been a land-grabber along the South Coast within historic times. The reverse has been the case. Along the shores of Kent and Dorset the prevailing westerly drift of wind and tidal current have piled up huge banks of shingle at Dungeness and Portland which extend our territory annually towards the south while the whole coast of Cornwall can show raised beaches (from another cause).

But space presses and we must get back to the Bay of Bays, resisting temptation to loiter where temptation offers the most seductive lure. Dartmouth has its surprises of narrow entrance and deep water within its land-locked harbour, and surprises of beauty as you follow the Dart estuary up to Totnes. I will not begin making comparisons. But I have always found that if I have met a man of Devon living far away from his own county you have only to mention Dartmouth to him to have his sympathies warmed up in an instant. Tor Bay sheltered round the steep, bold cliff of Berry Head is built of red sandstone, above which, from seaward, appear the blue heights of Dartmoor. The scene produced an exclamation of praise from the greatest enemy we have ever had, Napoleon Bonaparte, when he came anchor there in the *Bellerophon*. Here is Brixham whose fishermen are credited with the invention of the trawl net and even the discovery of the far-off Dogger Bank as a fishing-ground. All honour to them for their sustained boycott of the steam trawler! Paignton with its fine beach is in the middle of the bay. Torquay lies within the upper horn.

Round the bold promontory of Hook's Nose which has its secretive nooks crammed with luxuriant vegetation, such as Anstey's Cove, the shore goes north to Teignmouth, Dawlish and Exmouth. Here the cliffs are bright red and tend to form "stacks"—pinnacles of rock which stand isolated on the beach, a sport for the billows. Teignmouth was visited by the galleys of the King of France when James the Second had been driven to ask for his assistance. They landed and burned the

70 DARTMOUTH, DEVON: looking to the Harbour Bar

71 ODDICOMBE BEACH AMONG THE SOUTH DEVON CLIFFS NEAR TORQUAY

72 THE SOUTH DEVON COAST AT BABBACOMBE

13 A BRIXHAM TRAWLER

whole place down. A bright spot lightened the incident as the ships were leaving. Two galley-slaves on the ship nearest in shore managed to get on deck, jump into the sea, and swim safely to land. The bay curves round by Budleigh Salterton, at the mouth of the Otter, near Sir Walter Raleigh's birthplace. Beyond the shingle-barred outlet of the Otter and its dell cleft into the land lies Sidmouth in the midst of a fine sweep of coast, where cliffs appear more like a range of hills that have been pared vertically by the sea exposing their red sandstone cores.

At the end of this sweep lies Beer Head which is of quite a different order of rock, being our westernmost cliff of chalk. If you count the periods of geology as the coast moves east Beer appears to have come before its turn which should not begin till Portland Bill is passed. Its presence, however, is owing to a physical and not a geological freak. The correct order is followed if you skip Beer and go on to Lyme Regis where the cliffs change dramatically to a lurid blue—the lias. Lyme Regis has managed to conserve its old-world flavour much more intact than most places of a similar character whose popularity has exposed them to the trials and temptations of the summer season trade. Its old quay wall does not appear to have been materially altered since 1685, when the Duke of Monmouth landed on it to ring up the curtain on the awful tragedy of Sedgemoor.

The colourful formations of the cliffs from here to Weymouth are of particular interest as they form the termini of rock masses which go in regular belts across half England, reappearing, in due course, as cliffs on the North Sea coast. But that is not all, for these belts have more powerful interests for us than the colours of their rock or their specimen value in geology. They have undoubtedly modified both our artistic and political development. The lias belt is succeeded by that of the oolite which, in turn, is succeeded by that of the chalk. I will not say that the chalk has done a great deal for us beyond providing us with downland scenery, South Downs, Chilterns, Gog Magogs, in providing our agriculturalists with sheepwalks, our chair-makers with beech woods and our Londoners with artesian wells. But to the oolite belt we owe the majority of our fine churches, for the material has made the craftsman, and to the lias belt we owe the creative impules which threw us into the concrete age. This is the raw material of Portland cement. If necessity is the mother of invention, raw material is certainly the father of it.

The last of the chalk is seen in Flamborough Head and the lias and oolite end simultaneously in the adjacent cliffs of Yorkshire. Whereby hangs a strange tale, and I cannot pass by Lyme Regis without telling it. The lias is prolific of the great spiral fossil called amonite. Nothing shows the incurious nature of our ancestors so much as their apathy towards fossils. Explanations were offered from time to time, certainly, but the main bodies of philosophy and theology do not seem to have been touched by the presence of such curious anomalies as cockle shells in the heart of the rocks, said to have been created on the Second Day (whereas the cockle ought not to have appeared till the Third). Only the people of Whitby were wholeheartedly thrilled by the presence of their local fossil, the amonite. They had it placed on the coat of arms of their abbey (the town has borne it ever since). The amonite they held to be incomplete or rejected works of the Creator when He was making snakes. To help prove this theory, the upholders of it (in secret) carved on the amonites the reptilian heads which the Creator had failed to supply. Thus completed, the amonite was called a *snake-stone* and was said to be exceedingly lucky to possess. The rumour of amonite joss spread like similar rumours in our time concerning New Zealand jade, and a trade of manufacturing snake-stones (still kept very secret) grew up at Whitby and flourished to such an extent that the native supply of large amonites ran out. The deficit seems to have coincided with the dawn of the seaside visitor era. Naturally, supplies of the raw material had to be got from elsewhere. So ships came round the coast to the other end of the lias belt at Lyme Regis. Thus, for a time, Lyme Regis shared in a trade boom brought about (as it was still thought) by an aberration on the part of the Almighty.

Above the blue cliffs of Lyme Regis rises the hill called Golden Cap. Its yellow top belongs to the Cretaceous Period. The country behind Golden Cap is wild and interesting, and remains undiscovered by the tourist. Charmouth has a liasic background. In certain lights its cliffs are a vivid and spectacular blue.

Eastward thence begins the long barrier of piled shingle known as the Chesil Beach. The name (which is cognate with chisel) indicates flint. The barrier, which is one of the

74 THE HARBOUR AT LYME REGIS, DORSET

78 WEST BAY, BRIDPORT, DORSET: One of the Wooden Piers and the Cliffs of Burton Bradstock

Seven Wonders of the Five Seas, has a simple grandeur that can hardly be grasped by one who has not visited it. Its contrasts and harmonies, blended into a joint monotony for ear and eye to dwell on, achieve the same sort of unity of effect as that of a choir chanting Gregorian in a Norman church. Those endless grey-gold stones of hard, sharp flint all ground round! That endless assault by billows supple and friable! The endless crash and gnash and drawl!

Landward, the Chesil beach dams the outlet of streams, with the result that a narrow mere eight miles long, called the Fleet is formed. At the marshy head of this is the royal swannery of Abbotsbury where the King's birds are brought during the breeding season to nest. In sailing-ship days, in sou'westerly weather, if you got embayed along the Chesil beach nothing could save you, from which reason the sailors called this stretch Deadman's Bay. It does not end with the sudden turn in the coastline, but goes steadily forward with its scarcely perceptible arc till it reaches the Isle of Portland, which it serves as an exceedingly useful causeway. On a map, the Isle appears like the skull of an ictheosaurus to which the Chesil Beach fits as vertebrae in just the right place, while the saurian's beak is called Portland Bill.

The Isle of Portland is wedge-like in outline rising from the Bill to achieve cliffs nearly five hundred feet in height. The white oolite stone which makes them was popularised by Sir Christopher Wren (after he had built St. Paul's with it) as the most suitable material wherein to express Renaissance architecture. The industry of quarrying was jealously guarded by the Portland Islanders. How, living alone with their craft, they became detached from the world, while providing it with a principal medium for expressions of culture, is told by Thomas Hardy.

Within the causeway peninsula a fleet can lie at anchor in deep water, Nature's shelter from the west, south, and north having been completed by man to the eastward with breakwaters. From this large harbour whose extent is three square miles, the land trends northward to a point where a wide bay begins to swing the shore round to the eastward again. Here, the little River Wey forms a lagoon at its mouth called Radipole Lake, and Weymouth has sprung up at the entrance. George III brought the fashion to Weymouth. Outside the town he is commemorated by a large cut-out figure on the chalk hillside trotting jauntily off on horseback. Weymouth is still very Georgian, and the large express steamers of the Great Western Railway (which serve the Channel Islands) lying moored in the inner harbour, whose buildings are made as a setting for small sailing craft, give the measure graphically of how standards of life at sea have altered.

The chalk downs of the heart of Dorset reach the coast at this point and form cliffs of which White Nose Point is the highest, reaching 554 feet. Of the many charming bays and inlets along this sector, two are particularly remarkable. Lulworth Cove is perhaps Nature's prettiest attempt to make a harbour for man. Warbarrow Bay has, as centre-piece, to dominate its fan-shaped beach of fine sand, an imposing down crested with the rings of an Iron Age fort. But the encroaching sea has sliced it, from entrenchment to base, exposing its heart of chalk. The bay is close by another hill, Warbarrow Tout. Having a harder heart of brown contorted oolite, it has been carved into a spectacular sphinx-head and left (for the moment) as an isthmus.

The harder rocks now assert themselves and the Isle of Purbeck stands out into the sea with its corner-stone at St. Alban's Head. This isle is not detached from the mainland. Its only claim to insularity rests on being bounded by the lowest stretch of the Frome and a small tributary which rises to join that river in the high ground just behind Warbarrow Bay. Nor, according to the erudite, should we say St. Alban's Head but St. Aldhelm's. It is the sailors who have made the change, as in so many cases, preferring an easy malaprop to an academic mouthful. All the same, it is fitting that St. Aldhelm should be commemorated at the foot of the Purbeck Hills, for he was an authority on the building stones of the oolite belt. Did he not tell some workmen when he was out riding near Bath that if they dug in the earth at that particular spot they would find treasure in the earth? The treasure he alluded to was a kind of oolite which has been known down to this day as St. Aldhelm's Box (meaning treasure chest). But the building-stone which the Church has treasured more than all other comes from the Purbeck hills, namely a dark greenish-black shelly limestone which takes a high polish and is known as Purbeck marble. In the thirteenth century it was sent by sea to all the great churches in the kingdom for making the slender shafts which the master-masons of that day rejoiced in. St. Patrick's Cathedral in Dublin is wholly built of Dorset stone and its Purbeck shafts are made up in short sections

79 SWANAGE BAY AND THE OLD HARRY ROCKS DORSET

80 LOOKING TO BOURNEMOUTH FROM BRANKSOME CHINE

as specially adapted for export over seas in thirteenth-century bottoms.

Alas! how I am drawn into gossip when I promised an abridgement. But the Isle of Purbeck is holy ground in more ways than one. In coast and hinterland it is unique in all Britain. Rounding the next corner at Anvil Point, and passing the entrance to Swanage, guarded by two remarkable chalk stacks called Old Harry and his Wife, the coast rounds north into Bournemouth Bay. At the crook of it the whole Jurassic and Cretaceous Periods are changed for that of the Eocene whose first demonstration is the very interesting little inland sea called Poole Harbour. Beyond that the genius of this Period (which produced the London clay) gives us a coastline of sand-cliff prolific of pine and heath, carved at intervals into luxuriant ravines called *chines*. Bournemouth is set in the heart of this scenery. Its architecturally beautiful neighbour, Christchurch, has a replica in miniature of Poole Harbour of which the Roman galleys made use. Just here the even tenor of the shore is interrupted by the brave accent of Hengistbury Head, a promontory of reddish ironstone.

Now the Isle of Wight stands into the sea sheltering the broad, tranquil entrances which lead to Southampton. To the westward it confronts the seaman with the most dramatic challenge in all our coasts. From under an immense rampart of white cliff three sharp pinnacles of chalk called the Needles advance into the waves. And it can hardly be said that the lighthouse, perched on the last spire has not added to the beauty and grandeur of this approach. The island is a happy epitome of downland, cultivation and woodland. It has much that is individual. Its rainbow-coloured cliffs at Alum Bay; its white cliffs of Freshwater and the eastern horn of Sandown Bay; its sharp brown and black, sea-scarred cliffs of Chale and Black Gang which culminate in the solid grey rock mass where Ventnor is built—a town of verandahed houses rising tier upon tier in the eye of the sun from a red strand to an open mantling of woods, have not their like elsewhere.

Southampton Water is approached from the west by a strait called the Solent and from the east by another called Spithead. It runs inland along the fringe of the New Forest. This seaport has had a truly remarkable career. In many ways its history is twin to that of Bristol. It was the home of a powerful merchant guild and it was early nurtured on the commerce of the Mediterranean. At one time it held a monopoly of all sweet wines coming from that quarter in foreign ships, a gift presented to it by Mary Tudor in the fullness of her heart, when Philip of Spain came up Southampton Water to claim his bride. It had an admiralty all of its own whose badge, a silver oar, was carried proudly at mayoral functions. But its new achievement of attracting the large Atlantic liners from Liverpool is both the most remarkable and the most spectacular in its history. Chief among the answers to reasons for Southampton's recent successes will be found the one outstanding fact in which she is unique as a British port. Her waterfront is composed almost entirely of open quays. When other ports have been adding dock after dock she has concentrated on dredging fairway and berths and adding more and more open quays, so that a ship of any size can come alongside and tie up at any state of the tide without having to wait for gates on each side of a dock basin to be opened and closed. They say that trade is drifting south. Be that as it may, the attraction for it to do so in shipping facilities at Southampton are certainly increasing.

I fear that, for reasons given at the beginning of the chapter, I cannot do justice to the interesting coastline of the Isle of Wight. I must pass on up Channel with one more brief notice. Coming down the coast from the eastward, the top of the island rises over the sea-line like the land-fall of an enchanted country, a faint blue wedge, a gleam of dazzling white, and then an azure peak. These are respectively Boniface Down over Ventnor, the verge of Culver Cliff on the eastern horn of Sandown Bay, and Ashey Down behind Ryde. Ryde is on the Spithead Channel. Across the way, on the mainland, are three successive openings which lead into little inland seas of the Poole variety. But they have not the beauty of Poole as they are not fringed by high ground but by levels of mud, marsh, sand and gravel.

These lagoons bear the names of Portsmouth, Langstone, and Chichester Harbours. The last two communicate by a channel at the back of Hayling Island. The great naval dockyard is at the mouth of Portsmouth Harbour, but itself is not the port referred to in its name, which belongs to Porchester at the head of the inlet. Here the Romans placed the first of a great system of fortresses to defend the country against raids of the Saxons, anticipated as early as the beginning of the fourth century. Porchester remains almost intact, as does the last but one of the chain—Borough Castle, near

81 A CHALK LANDMARK FOR CHANNEL TRAFFIC: the Needles, Isle of Wight, from the Air

82 SUNSET OVER THE SOLENT: a Cowes Week Scene, with the Royal Yacht on the Right

3 SOUTHAMPTON DOCKS FROM THE AIR

84 ENTERING PORTSMOUTH HARBOUR: the ~~Helmsman~~ Leadsman at Work

ERRATUM

Fig. 84, Title, *for* "Helmsman" *read* "Leadsman"

Yarmouth. This trinity of natural harbours though it lacks scenic grandeur has all the charms of the mud-muse which I have dwelt on more fully in my account of the Thames Estuary.

Selsey Bill is the first promontory of Sussex. It is as flat as a pancake but, in the distance, the South Downs can be seen like a host of clouds bowling seawards. These are not mountains, for they do not rise anywhere above eight hundred feet. But, as hills, they are wonderful examples of how fatuous it is to judge scenery on a basis of size alone. Their shapes and proportions give them a kind of grandeur on misty days that you might search for in the Pyrenees and the Rockies and not find, and perhaps the sea is never felt to be of such a tender blue as when, while you watch it, your other senses are engaged by the song of the lark, the pressure of the turf and the smell of the thyme on these downs. This range has passes of an unusual kind, for the streams do not rise in it, they cut gaps right through it. This is quite contrary to Nature. One's surprise is made complete by noticing that the same thing happens also in the North Downs—only in an opposite direction. The answer to the riddle is that a mountain watershed of a younger geological formation has existed between the two ridges on the site of what is now the Weald of Sussex and Kent. Erosion has carried the mountain off the map but the ravines remain, and they add greatly to the view of the Downs from the sea.

Thus the highest downland mountain, Chanctonbury Ring, which stands behind Worthing, has the extra good fortune (like Snowdon) to have its individuality accentuated by passes cleft boldly either side of it. Through these gaps run the rivers Arun and Adur. The former comes down to the sea at Littlehampton, passing the great castle of the Dukes of Norfolk at Arundel. The latter forms the harbour at Shoreham, between Worthing and Brighton, where a whole bungalow town has risen within recent years. A third of these curious corridors opens in the Downs at Lewes to let the Ouse through, whose source is in Ashdown Forest. It goes to sea through the cliff wall which rises east of Brighton, having carved out the sheltered cross-Channel port of Newhaven. The last of the series is the Cuckmere. Its joint conspiracy with the sea is responsible for the fine cliff of Seaford Head.

The next inroad on the Downs is made by the sea itself.

At Beachy Head they have been fairly rounded up and cut off. If you stand on Seaford Head and look towards its rival—which is nearly twice its height—the cliff contours are seen to cut a caper in seven waves of undulation which has gained for them the inevitable name of the Seven Sisters. This is an actual cross-section of the last South Down in process of being eaten away by the sea, and Beachy Head is its five hundred-foot summit. So, in rounding Beachy Head, we cross the South Downs on the sea-level, and on the farther slope find Eastbourne snugly tucked away between the Downs, the Weald and the sea. Hencforth the low cliffs (where they exist) will be of sand (the Hastings sands of geology). We shall not see the chalk again till we find the North Downs being lopped off in the same drastic manner.

This is the coast which cradled our naval power. Before the regular establishment of a royal navy the King relied chiefly on the services of the Cinque Ports—Hastings, Romney, Hythe, Dover and Sandwich. The arrangement was that, in exchange for charters freeing them from tolls, dues and other burdens, these ports would provide as follows on demand; 57 ships with 1,197 men and boys for service at sea during the space of fifteen days in any one year. They were to find themselves in all gear and rations for this period, and the time counted from the moment the fleet weighed anchor. For service beyond the full fifteen days they were permitted to put in a bill for overtime. The Cinque Ports fleet was used principally for raids on the Continent, but it played a leading role in the conquest of Wales and not infrequently its keels rode the waters of the Solway and the Tweed. The Cinque Ports alliance was augmented by certain other ports distinguished as the Ancient Towns. They included Pevensey, Rye, Winchelsea, Lydd, Folkestone, Deal and Walmer. The whole of this powerful maritime community was ruled by the barons (that is, burgesses) of the Cinque Ports under a Lord Warden.

The connection of the Cinque Ports and Ancient Towns with the Yarmouth Fair I have mentioned at the appropriate place. Of the elements which built up the power and prestige of Britain this curious amphibious mobile palatinate was surely one of the principal foundation-pieces, and the story of its decay is one of the saddest. Yarmouth, whose fair must have provided the barons with a handsome steady revenue, became independent, while the sea not only silted

85 SUNSET OVER BROADSTAIRS

86 SPRING WEATHER AT BRIGHTON

87 RYE, SUSSEX, FROM THE OLD HARBOUR

the ports but built up the coast beyond many of them, converting them to inland townships. The principal damage was done in the neighbourhood of the promontory which succeeds Beachy Head, namely Dungeness. This is a bud of land which still grows, year by year, at the rate of about one foot per annum, as favourable conditions of wind and tide continue to pile it up with flint shingle transported from the ruin of the South Downs.

To-day the hinterland behind Dungeness is a vast sheep-walk called Romney Marsh. From being the hub of British enterprise this region became so detached from the stream of life that fifty years ago people were said to believe in the existence of six continents in the world—Europe, Asia, Africa, America, Australia and Romney Marsh. At any rate I remember it, half that time ago, as having a seal of peacefulness and far-offness more impressive than that of a cloister. I have not visited it since it has been equipped with a real scale model railway run on commercial lines, but I should imagine that such a happy and child-like solution of its communications would only enhance the sixth continental sense of other-worldliness.

The North Downs reach the coast at Folkestone where their mass has been carved to a sheer wall of chalk cliff which extends to the South Foreland broken only by a low interval at Dover. Abbot's Cliff and Shakespeare's Cliff stand between the two famous cross-Channel ports. Dover is, in itself, a large theme which must remain outside the scope of my book. We must pass it by merely noting that it preserves one of our most interesting Roman remains, namely that of a lighthouse. This relic has looked out on many strange adventures in arms embarked upon the narrow Strait, between the passing out of the Roman legions and feats of the Dover Patrol, and I am tempted to touch on one which shows that chemical warfare was not unknown to our Norman ancestors. It happened the year after King John died while civil war still endured between the royal power and the rebel barons. The barons had managed to get help from the French. After the Battle of Lincoln, however, things went badly for their party, and a strong reinforcement set off from France for the East Coast. That grand old warrior, Hubert de Burgh, was at Dover and the Cinque Port fleet was in the harbour. They left their moorings, came out and got to windward of the enemy as he passed up Channel. Bearing down, they let fly not only a swarm of arrows but also a cloud of quick-lime, shaken on the breeze out of bags. The victory was a decisive one and an account of it winds up by saying: "The Cinque Port seamen returned in triumph towing their prizes, after throwing the common soldiers overboard, and taking the knights to ransome."

The narrows between the coasts of England and France bears a subtle difference in nomenclature in map and chart. The latter refers to it as the Dover Strait, the former as the Straits of Dover. It would be interesting to discover when this hair was first split. The two principal places within the Strait are Deal and Ramsgate, both of which owe their rise and first prosperity (before the coming of the tripper) to the presence of what might be called the most important sea waiting-room in the whole world—the roadstead known as the Downs.

In sailing ship days the position of the Straits of Dover was a very crucial one. If you came out of the port of London with a fair wind the same would be a foul one for running down the channel, whereas a fair wind up Channel would be foul for entering the Thames estuary. A sheltered anchorage at the junction of the courses, where ships might wait for a veer in the wind was essential. Such an anchorage was found in the Downs which seemed to have been specially arranged by Nature for the benefit of the mariner. But there was a reverse side to the picture. The nature of this shelter depended on the position of certain treacherous sandbanks known as the Goodwins. While they broke the force of the seas on the roadstead they threatened all shipping going in and out of it, and were prolific in quicksands.

The Deal boatmen were a nimble, hardy, dare-devil race who lived equally on the evil reputation of the Goodwin Sands and the boon of the Downs roadstead. They enjoyed a world-wide reputation under a curious nickname—the *hovellers*. They acted as pilots of the intricate and ever-shifting Goodwin, as life-boat and salvage men in emergencies, which were frequent enough, and as carriers of provisions and letters between the ships and the shore. They faced the worst weather in their open luggers which were pulled up on a steep bank of shingle and could be launched by the aid of gravity and rollers in an instant.

Ramsgate became an important place when its harbour was built by Smeaton for the Government as a free haven of

88 SHAKESPEARE'S CLIFF AT DOVER

89 BEACHY HEAD, SUSSEX: the Chalk Cliffs and Lighthouse

90 LOOKING TO FORENESS POINT FROM THE SANDS OF CLIFTONVILLE, KENT

91 REGENCY HASTINGS, with the Castle Hill behind

refuge for smaller craft to save them from the necessity of lying in the perilous roadstead among the Goodwins, and to secure them from the expensive attentions of the hovellers. The stationing of steam tugs and a lifeboat at Ramsgate was the first serious blow to the trade of the Deal luggermen. To-day the hovellers are extinct—all glory to their memory as England's most intrepid seamen! Steamers, unmindful of the winds, have no need to pause in the Downs, and the way through the Goodwins is found for them by Trinity pilots who are guided by an efficient system of beacons, unknown in the old days.

But the old terror of the Goodwins is well within my own memory. I recall as a boy, in the nineties, how Ramsgate during the winter months was in a constant ferment of excitement. How, when the wind sprang, the waterfront was crowded with people straining their eyes towards the south. How the crash of the rockets brought the whole town out. How the harbour rang with cheering when the old paddle-tug returning hung perilously battling in the harbour mouth, towing some forlorn ship whose sails were blown to ribbons, or whose masts had gone by the board, leaving a raff of wreckage on the deck, about which a cowed, starved crew still clung.

CHAPTER VI

THE NORTH SEA

Until the Great War, in which the name of Germany lost favour with us, this sea was always known by two names on our maps and in our geography books. It was called the North Sea or German Ocean. In the seventeenth and eighteenth centuries it bore the latter name only. This, on the earlier maps was Latinised as Oceanus Germanicus. You may look to the classical renaissance for the adoption of this name, but I fancy its use was really popularised through the famous doings of the Hanseatic League. The League had established centres of commerce, called factories, in London, Ipswich, Norwich, Yarmouth, Boston, York and Hull. They were much like our merchant houses in the treaty ports of the East with the interesting difference that the representatives of Hansa which occupied them lived under a rule of life and conduct nearly as strict as that in a monastery. It was not a religious rule but a commercial one. It enjoined among its members not only strict temperance, but also a rule of celibacy which forbade them either to marry or to take mistresses from the nation in which the factory was stationed. The factories are clean gone, the name of Hansa is forgotten, and now, having given up the name German Ocean, we have cast off the last link which bound our memories to the fact that in the Middle Ages the East Coast was the commercial coast of England.

But the name North Sea is not new. It was in use by sailors before the other, for distinguishing between the Baltic (the East Sea), the Atlantic (the West Sea) and, latterly, the Pacific (the South Sea). I fancy that sailors have never given up the first and last two of these ancient styles for the more fanciful titles of the cartographers. For it is still a mark of sea breeding to speak of the Western Ocean instead of the Atlantic.

I have some regrets that we have lost the name German Ocean, for it imparted a certain flavour of foreignness and a sense of wide dimension that are lacking in its present style. One needs something of a reminder that the East Coast has for her sister shores those of five countries (not counting the seaboard of the Baltic), that the flying foam of this turbulent expanse is the broth of the great Nordic Melting-pot,

92 GRAVESEND, KENT: Gusty Weather on the Thames Estuary

93 LOW TIDE ON THE LONDON RIVER AT BARKING

94 ROCHESTER AND THE MEDWAY

95 MEDWAY BARGES AT ROCHESTER, KENT

from which Angle, Saxon, Jute, Fleming, Dane, Norseman and Norman have come to be fused in one Englishry.

Apart from history, its individuality, as compared with that elusive element in our other seas, is striking. The tide is peculiar. In the way its inundation from the Atlantic, round the north-east of Scotland, fits in with that through the English Channel, it enables a steamer to make use of a one-way current for more than twelve hours at a stretch. The flood tide runs southward down the coast and there is seven hours difference between the time of high water at the mouth of the Humber and the Straits of Dover. So that if a ship proceeding from Hull to Zeebrugge choses her time, she can get the tidal stream with her all the way there and, likewise, have it all the way back; which means a considerable saving in coal for the owners. The atmosphere, both in the actual and in the poetical sense, is very different from that on the other coasts. The north-easter is the prevailing wind. It is a land of perishing winters, bracing springs, torrid summers, and amazingly colourful autumns, especially in the fenlands and the broads. And you can see a rainbow over the sea at sunset— which you cannot do on the other coasts. As to the poetic kind of atmosphere, it is conveyed through many subtle hints to eye and ear—the red-tiled roofs, the distinctive fishing-craft of Northumbria and Yorkshire, the churches (unrivalled in their magnificence), the ubiquitous fig-tree, and the local accent and idiom of East Anglia. And the folklore of this part is augmented equally and oppositely by the losses suffered through coast erosion, and also by the amazing finds to be had on the seashore, amber, jet, cornelian, agate, bones of primeval elephants, stone coffins, mediaeval and Saxon jewellery.

These losses and finds have fostered a tradition of church bells heard ringing under the waves, of cities seen in the deeps on still days, with dim figures moving in their seaweed-tangled streets, of bearded hairy mermen caught by fishermen in their nets. And then there are the tales of the wrecks. The East Coast easily holds the record for sea disasters, for there was no safety for the sailing-ship in bad weather between the Thames and the Humber, and between the Humber and the Firth of Forth (it was usually suicidal to try and enter Whitby or to attempt to cross the bar of the Tyne). In fact, wherever you go along this coast there is a sort of reserve of beautiful melancholy, having its source in a threat or a vision out of the sea.

The North Sea has a magnificent entrance. It may not be so spectacular as that of the Mediterranean with its Pillars of Hercules. But, to the Englishman, returning home from distant lands, the vision of the North Foreland on a summer morning gleaming through that soft haze which perennially besets the Dover Strait, must seem the greatest miracle achieved by any coastline in all the world. As he breathes the honey-dew which wafts subtly from the fields of lucerne on the cliff-top he will taste a sort of essence of the sweets of scenery said to be peculiar to his native land. And they may, as it were, set the spectacle of that great chaste, white corner-stone of Albion to music, so that it appears something more than a freak of nature, a hewn emblem shaping out a wistful ideal.

Indeed there is something mystical about the anatomy of the whole of this heel which "spurns back the ocean's roaring tides." From Margate, that time-honoured bourne of the London tripper (since the days when the sailing hoys first dropped down the Thames with them) the coastline decreases in height till it reaches a country which is all flat moated pastures and huge skyscape. Through the midst, a trivial watercourse finds its way to the sea at Reculver. It is the remains of the estuary of the Wantsume, formerly a navigable sea-arm which insulated the Isle of Thanet on the north-west from the rest of Kent. The Romans planted their fort, Regulbium, at the narrows, well away from the sea-mouth of the estuary. But the site is now being attacked by the waves and for more than a hundred years has been shored up with walls and protected with groins. The place is marked by two towers called by the sailors the Reculvers. Actually, the name is singular. For this is the former parish church of Reculver. As the sea encroached, the congregation receded till, in the early nineteenth century, the nearest parishioner lived a mile away, and the church would have been wholly dismantled for transportation inland had it not been that the twin western towers were an invaluable mark for sailors. So the work of destruction was arrested by none other than Trinity House, who patched up what was left of the building and made the sea-wall and groins. In doing this they preserved to a more appreciative age not only half the Roman fort of Regulbium but one of the most interesting churches in existence. The towers and outer walls of the nave are Norman. But, within this shell are the foundations and part of the walls of a Saxon

96 AN OLD TIDE-MILL AT ST. OSYTH'S, ESSEX

97 RECULVER CHURCH, KENT: once within an Estuary, now on the Open Sea

98 DUNWICH CHURCH, SUFFOLK: a photograph taken in 1922, showing Nave and half Chancel standing. Now all is gone

church that were raised in the time of Archbishop Theodore who came to Canterbury in the year 669. Thus, the Romans built these forts with the express purpose of keeping the Saxons out of Britain. Yet it owes its preservation to the presence of a Saxon work. It is one of those situations which the muse of history enjoys.

The sea marshes of the Wantsume, fenced from the briny element and made prolific of good, sweet grass, are less wild than Sedgemoor and much less eerie than the Sands o' Dee and the Solway. But they have a mystic element of old loneliness about them that makes them comparable to no other place except Romney Marsh and the rest of the great eastern gulf of which they are a part—the Thames estuary.

London is the greatest and also the most unostentatious port in the world. In this way it is like the heart of the best kind of Englishman. But the Thames estuary is positively secretive. You may compare it with the Japanese proverb which says that the darkest part of the room is often at the base of the candlestick. How many Londoners who race all over Britain to find novelty of scene know the subtle charms of their own estuary? The traveller by sea arrives in the London River without more than the barest glimpse of the great estuary. From the Forelands he is conducted to the Nore by a chain of lightships. But even if he were to see the land, he would make nothing of it.

Near Reculver is Herne Bay, actually not a bay at all but a promontory in the making. The name is a monument to coast erosion. Whitstable shares with its opposite number on the other side of the estuary, Colchester, the chief honours of bedding the oyster—as purveyed to their majesties the Emperors of Ancient Rome. Whitstable is at the mouth of the Swale, a strait which is more like a salt-water river of the kind that the Niger sprouts at its mouth. If you follow it, it will take you by devious reaches into the mouth of its parent stream, the Medway. On its last loop the cross-channel port of Queenborough is built. The Swale isolates Sheppey from the mainland with the isles of Elmerly and Harty. The horrors of running aground are, in the Swale (at its Whitstable end) much aggravated by the penalties attaching to skippers who pile up their vessels on oyster-beds.

These Thames estuary muds have other honours besides that of supporting Whitstable and Colchester natives. They have a beauty. Whistler nearly succeeded in interpreting it—

but not quite. It is of such a subtle kind that I fancy it will never submit to a one-dimension art. It is bound up with shadow profiles in moonlight multiplying horizons above the sheen of tide-dimpled water, with swiftly moving skeins of flying sea-fowl, with the thrilling dawn-cry of the curlew and the sad vesper notes of wading birds, with the sounds of guzzling from the zostrea beds where wild ducks hold their midnight banquet.

The forlornness of sea muds is superb. Such commonplace objects as empty barrels and broomheads mounted on crooked stakes to mark the channel acquire poetic distinction of a high order. The loneliness of a Sahara is static and daunting. But this loneliness is dynamic and stimulating.

The mud-ography of the Thames estuary is a recondite life-study, known only to the captains of London's famous fleet of sailing barges. The lower reaches of the Medway have the most intricate shores. They are jigsawed into creeks and nesses—all named. These, in turn, are jigsawed into saltings and spits—all named too. Every generation seems to have left its labels there. Perhaps Bishop's Ooze was called after Odo of Bayeux who built Rochester Castle in the eleventh century. Who named Humble Bee Creek and Tea Pot Hall?

The reaches of the Medway lead up to Chatham and Rochester, an old city which seems never to have grown older so well has it preserved the flavour of its early days. The river is navigable as far as Tonbridge, mounting by one lock to Maidstone and thirteen thereafter. On its gentle passage to the sea it has carved itself a way through two ranges of the chalk downs. From its breadths and its narrows rise shores of corn-land and pasture, brake and woodland, while church towers, mansions, farms and cowled oast houses grace the landscape at every bend, I don't know where King Arthur's Camelot was, but Tennyson's Lady of Shalot can have drifted down no other river than the Medway.

Just to seaward of the Nore lightship lies the boundary of the Port of London Authority, a body conjured into being in 1909 by the adroit management of Mr. Lloyd George. What the port of London *is* you may gain a vague idea of from an aeroplane. To know it in detail probably requires the dedication of a lifetime. The greatness of London as a port does not, as in the case of our other ports, depend primarily on local trade and manufactures but on the fact that the City is a world market. The great warehouses of wool, of ivory and spices, the cool, dark cellars of the bond stores where pipes of rare wines mature in a sumptuous silence cannot be seen elsewhere. Yet London in 1800 had only one dock, and that was a private affair. The ships lay in the river and the loading and unloading was all done by lighters who "worked in" with the owners of wharfs. The remarkable fact is that in spite of the creation of the present vast system of docks, the wharfingers and lightermen have never lost their ancient privilege of "free water." Here at least is a clear case of vested interests of the most powerful kind failing to infringe the liberty of the subject. Although the docks relied for their revenue as much on the freights as the hulls of the ships which came within their gates, they were powerless to prevent the lightermen entering the enclosed water and relieving the ships of their burdens at their own terms. The fact that the wharfingers were able to preserve their rights has made the waterfront of London unique in appearance, as it has preserved the old amongst the new. Dropping down the river from the Pool you will see an endless range of wharfs, stairs and other landing-places, many of which have the same backgrounds that confronted the tall ships of sailing days, the same quaintly perched houses, inns and shops, the same stairs at Barking where Judge Jeffreys met his well-merited doom. Along the winding corridor of the world's greatest port, show and display of any kind are the last things in evidence. Factory, power plant, distillery raise their shapes undisguised. And the palace of Greenwich with its regal Palladian façade accepts them as good and useful neighbours, as worthy of London as the courts of kings.

The north side of the Thames estuary has an even ampler share of mud-flats, of islands and creeks. The Essex marshes with their great farmsteads and their long vistas of sea and sky have the same attractive spirit of *old* loneliness found at Reculver—with strong local differences. This country is often called a Dutch landscape, and the name implies more than a mere resemblance to the dyked pastures of the Netherlands. Actual Dutch settlements have been made here, as at Canvey Island in 1622 when Joas Coppenburgh an engineer from Holland reclaimed it for a reward of two thousand of its newly won acres. Other Dutchmen came later in the train of Vermuyden. If, at this date, we can find traces of these immigrants from the Low Countries in buildings, customs and names, it is more than likely that they found plenty of living links with a much larger migration from the Netherlands which had arrived a little more than two and a half centuries earlier—the Flemish weavers of the time of Edward the Third.

In the Middle Ages, an apt name for all East Anglia would have been the Wool Coast. From Saxon times until the fourteenth century the principal trade was that of exporting fleeces to the Continent, where the wool was made into cloth and re-imported to this country. It was Edward the Third who enacted the first drastic protection laws. He put an embargo on the export of fleeces. This rough and temporarily ruinous measure was none the less speedily effective. As the wool ceased to go to the weavers, the weavers came over to the wool. The success of this measure is reflected to-day in the great churches of Essex, Suffolk and Norfolk whose size and architecture tell of great populations, of great wealth and culture where these are no longer found in their neighbourhood.

Essex has a strong forest tradition, for the old royal forest of Essex (of which Epping is such a noble relic) was one of the last to be disafforested, owing to its convenient proximity to London. In scattered woods there are many intact survivals. The arboreal tradition is strong and extends even to places on the dead flat coast, where trees would not ordinarily be looked for. Constable sought and found some of his most successful expositions of English landscape in the Essex blend of elm-tree, homestead, water and water-meadow.

The great intricacy of the Essex creeks have prevented the spoliation of this lovely countryside by compelling main lines of road and rail to keep well inland and isolating the two large seaside resorts of Southend and Clacton from each other. Between them lie an amazing ramification of creeks and channels carving out peninsulas and islands which confound even the motorist explorer.

Eastward of Southend (meaning the south end of Prittlewell parish) lies Shoeburyness where the gunnery range is situated. The trial barrages from here have as their objective the Maplin sands which is perhaps our largest and loneliest seashore. At low tide this great sandbank exposes for twenty miles, with a maximum breadth of between six and seven. Beyond the Maplin a gulf opens which gives to the sea the rivers Roach, Crouch, Blackwater and Colne, together with the oyster-paved creeks which indent the coasts on either side of Mersea Island. From the villages of East and West Mersea from the picturesque quay of Brightlingsea and from Colchester

99 A HAUNT OF YACHTSMEN: the Anchor Inn at Maldon, Essex

100 THE QUAY AT WOODBRIDGE, SUFFOLK

101 BLAKENEY CHURCH, NORFOLK: the East Tower was built as a Beacon for Sailors in the Fifteenth Century

102 ORFORD CASTLE: a medieval Stronghold of the Suffolk Coast

oyster fishery is prosecuted, and a powerful fleet of smacks goes forth to drag the shrimp-infested gulf. Up the rivers sea-going craft reach Chelmsford, Malden and Colchester.

It seems a little odd that in a district where every conceivable physical feature in land and water is named, this gulf which cuts an appreciable figure even on a small scale map of England, should go untitled. The sailors are satisfied with naming the deep water which lies off it—the Wallet. It is a tacit acknowledgement of a no-man's-shore, an interval across which landsman and seaman can have no contact. Between the Wallet and the lesser gulf to the north-east called Harwich Harbour is a peninsula which has had a name though it is already so far gone in decay that it is not marked on large maps any longer. It is called the Hundred of Tendring. It confronts the sea with a line of low chalk cliffs and a sandy foreshore. Here are forgathered a trio of seaside resorts— Clacton, Frinton and Walton-on-the-Naze. The voyage to Clacton from Tower Pier is one of the Londoner's principal and cheapest outings. A solemn and impressive note in the marine approach to this citadel of gaiety is the sight of St. Osyth's Priory whose ruin is conspicuous, although it stands three miles inland. At the spoiling of the monasteries, Thomas Cromwell, the King's agent in that vast scheme of destruction, had earmarked St. Osyth's for his own uses. In spite of his power and position he had some difficulty in manœuvring it away from other pressing claimants. He succeeded. St. Osyth's, standing on a wooded rising ground above its quiet old village and the waters of the calm creek, has a charm of peace and detachment which is unusually striking, even for a monastic site. One can feel that a desire to end his stormy life in such a place may have been stronger than mere cupidity. But Cromwell's was the fate of Ahab who cast covetous eye on Naboth's vineyard. He had not been in possession of the place for a year before his head was struck off on Tower Hill.

There is no other landlocked harbour between Harwich and the Humber. Harwich has always been a premier port in the history of the English seas, and is constantly mentioned in Hakluyt's *Voyages of English Seamen*. It received the bulk of the Fleming emigration, mentioned above, and also the refugees from Alva's persecution in the Netherlands. Its connection with the Low countries is still a vital one. The L.N.E. Railway run regular express steamer services from here to Antwerp, Zeebrugge and the Hook of Holland, and from here the train ferry takes goods trains over bodily to the outport for Bruges.

The Stour comes down from Manningtree in an estuary which begins with dramatic suddenness, maintaining a width of over a mile from there down to Harwich. You may pass through a tidal lock near Manningtree and navigate the river a further twenty-eight miles (on an inland craft) to Sudbury, passing through Dedham whose water-mill formed the subject for one of Constable's most famous pictures. The companion estuary which opens into Harwich Harbour is that of the River Orwell. Fairly large steamers can use it up to Ipswich as the channel has a minimum depth of 17 feet.

At the foot of the tongue of land between the Orwell and the Deben estuaries is Felixstowe with its mile-and-a-half long pier. Woodbridge stands on hilly ground at the headwaters of the Deben estuary. Its quays lie in a delightful setting which seems made to receive the gaily painted hulls and immense tan sails of the London barges.

The turning-point of the coast, at which the Thames estuary ceases and the open sea begins, is Orford Ness. Between this point and the mouth of the Deben River is a slightly convex stretch called Hollesley Bay. Here there is deep water right up to the land. In the offing a large sandbank called the Whiting breaks the force of the sea. It is an anchorage where, at many critical junctures in history, large fleets have ridden. Probably for this reason there is a special concentration of martello towers here. No less than four, each within about a mile of the other, guard the approach to the Ore River. The Ore may be described as a salt-water river grafted on to a freshwater one as an afterthought of Nature. The River Alde, of many tributaries, comes down from the Suffolk broads in a reasonably direct meander to the sea at the back of Aldeburgh. Yet it fails to reach its objective by the distance of an easy stone's throw (across a shingle bank). Having failed, it instantly gives up the ghost. That is, it parts with its name (the possession of which rivers are more tenacious than any other known thing) and becomes the Ore. As the Ore it runs at right-angles to its old course and keeps nearly parallel with the sea for fifteen miles, until, in fact, our most emaciated peninsula breaks down in archipelago and lets the tenacious land-lubber out.

The delightful little village of Orford is situated about half-way along the stream which gives it its name. It has two splendid ruins that have truly been the sport of Time, for he has dealt with them in reverse measures. In the church he has desolated the fine Norman chancel and preserved only the nave of later date and part of the tower. In the castle he has preserved the whole of the great Norman keep practically intact, sweeping away the whole of the later buildings so that not one stone is left upon another. This keep, raised by Henry the Second in 1165, is one of our best and most intact relics of the Norman Period. Even the stone altar in its chapel remains (though the top is gone). A sketch made of the castle in 1602 shows a cresset fire burning on top of one of the turrets of the keep, indicating that from its earliest days it must have been used as a lighthouse for mariners.

From that turret you look out across the Ore to Orford Ness where the sea is five fathoms deep close to the shore. You also overlook the Ore's island of Havergate where Margaret Catchpole's cottage still stands—she was our most notorious woman smuggler. Turning your eyes inland you may see the woods of Staveton Park where immense aged trees survive as relics of the ancient forest. Our present Department of Forestry has chosen to plant a large area of fir-trees right alongside of it. At either of the inns in Orford you can get oysters all the year round. They are transported from the Mersea beds and kept alive in *vivaria* alongside the quay. During those months which are lucky enough to have an R in them you will get the Colchester native, in those which have not, the American blue-point. To eat them on a truly Jacobean (meaning W. W. Jacobs) waterfront at a house whose sign is The Jolly Sailor, in a bar parlour which has a glass case containing real stuffed dogs the size of mice, adds a rare flavour unobtainable in the most select city bar.

The Ore and the Alde give splendid scope to the calm-water yachtsman, and, for a long distance inland, the countryside is enlivened by visions of red and white sails over green pasture and golden grain. Aldeburgh and Southwold, the two more dignified seaside resorts of Suffolk, have much in common that is distinctive and charged with local character. Either by accident or by skill in council they have both managed to avoid the contamination of the tripper. To the casual visitor Southwold appears both the more interesting and the more democratic of the two. In its architecture there is a great deal more to remind one of the ancient days. Yet it has not the old-worldliness of Aldeburgh. That is an indefinable quality which seems to make an impression on the heart rather than the mind. And if you have it in a place, it will not be a real asset and warrant you keeping up your prices unless you have geography on your side as well.

Aldeburgh is built on two levels. The lower has for long been attacked by the sea. In fact just about half the town has been washed away, so that its moot hall which used to occupy a central position is now at the water's edge. This fact perhaps accounts partly for the air of charmed melancholy which is one of the amenities of the resort. Bathing operations are conducted from machines of the most portentous type, each having double cabins. They are lowered by gravity down the steep shingle upper-works of the beach and raised by capstans.

Between Aldeburgh and Southwold the coast is clean cut, the sea rising from a floor of sand to plunge on a glacis of shingle below cliffs of grey-yellow Pliocene crag. The main road lies away from the coast, which is only approached at most points by farm lanes. If the traveller on foot passes between these two towns, making choice from his ordnance map of the most faintly indicated roads, he will find himself traversing some of the most lovely small pleasuances in England. This is so especially when the corn is ripening and the heath is in bloom and the bullrush head is swelling under its velvet. On the great moors of the Pennines, of Wales, and Scotland it is the ling which predominates and gives a billberry-coloured stain to the heather masses. But the heath is the richest of the "heather mixtures," and here are great tracts of heath alone, making crimson acres threaded by yellow paths of sand. These alternate quickly with grainlands and thick woods, some of birch, some of oak. Old churches suddenly reveal themselves through hedgerows of elm. If you have seen a thousand fine sunsets you may yet witness an unrivalled one here.

Two-thirds of the way to Southwold is Dunwich. It is a bourne of sightseers who come not to see what is, but to muse on what has been. The Venerable Bede speaks of it as the Capital of East Anglia. That was in the seventh century. Much later it is spoken of as having a considerable settlement with a monastery, churches and a harbour. The harbour disappeared some time ago. In 1914 the cliff had crumbled up to the foundations of the chancel of All Saints' Church whose axis was at right-angles to the advancing precipice.

103 NORTH SEA DRIFTERS RETURNING TO YARMOUTH

104 FLOUR MILLS AND BARGES AT IPSWICH, SUFFOLK

105 A NORFOLK SEASIDE TOWN: Cromer from the Cliffs

In 1922 only half the nave was left. To-day it is all gone. The ruins of the tower lie on the beach awaiting the last touches of disintegration.

Dunwich beach is the most prolific hunting-ground for finds in all Britain. That gleaming bank of shingle teems with objects belonging to Celtic, Roman and mediaeval Britain. They appear, disappear and reappear after successive gales. The proprietor of the beach refreshment-hut has made a substantial collection and presented it to the local museum which has recently been established. Among the less distinguished objects which this man has picked up at regular intervals are Dutch wooden sabots. The odd thing about them is that they are all left foot. I will not attempt to solve this riddle.

In the past, the Dutch have kept us reminded of their proximity by other sorts of sabotage, herring-poaching, piracy, privateering and war. As to the last, the chief local incident was the Battle of Sole Bay fought in 1672 within earshot of Southwold between our fleet under the Duke of York (later, James the Second) and the Dutch fleet under De Ruyter. New York had only recently been named after our admiral (its former style was New Amsterdam), and its first British governor, Nicholls, was killed in this action of Sole Bay. You may still see the shot which killed him, placed over his tomb in Ampthill Church in the shire of Bedford. Sole Bay appears to be no longer marked on the map, which is, no doubt, another evidence of coastal erosion.

The law which governs the dispersal of relics is certainly past finding out. As far as I know, you will find no tangible remains at Southwold of the Battle of Sole Bay, though you will find there six guns taken from Johnnie Cope at Prestonpans by Bonnie Prince Charlie, which were after retaken by the Duke of Cumberland at Culloden and presented with a flourish to Southwold. But the Jacobites have the last laugh, for the guns mount guard over the outlook towards the site of the most useful victory ever won by a Stewart prince in person. Southwold has the amenities of a seaside resort unobtrusively combined with the quaintness of an old-fashioned country market-town. Its crowning glory is its church.

The next two rivers which come down to the sea are the Waveney and the Yare. Their conjunction with a sea roadstead and proximity to a famous fishing-ground are responsible for the two great herring ports at their mouths of Lowestoft and Yarmouth. Yarmouth is the more ancient and has always been the more powerful of the two. But Lowestoft has a respectable antiquity. Their rivalry has been more keen and bitter than that of any other neighbour ports on our coast, always heightened by the fact that they are in different counties which are, themselves, noted for their rivalry.

Perhaps no other industry, not even wool, has modified the history of north-west Europe so profoundly as that of the herring fishery. The power of the Hanseatic League was gravely shaken when in 1425 the herring suddenly changed its spawning grounds from the Baltic to the North Sea.

Lowestoft has a genius for the unexpected. To begin with, you do not pronounce its name as spelt, but leave out the *e* and the second *t*. The railway station, and not the parish church, is the centre of this large town. In fact the parish church stands nearly half a mile away from the haunts of men, right out in the open country. The reason for this is said to be that the builders took their cue from the ruination of Dunwich and built where they thought the sea could not get at it until the Day of Judgement—an event they looked for confidently within say a couple of centuries at the outside. In both expectations they were wrong. For the sea, behaving like the Weather Clerk when he sees you go out with an umbrella, made up its mind to recede from this part of the coast and attract the congregation with it as it went.

In the early nineteenth century when the canal and "navigations" mania was at its height Lowestoft aspired to be an outport of Norwich. They were advised by Telford to cut inland and tap the Yare, Yarmouth's river. Yarmouth was invited to join, but rejected the offer. After the usual difficulties powers were obtained from Parliament and after expending £140,000 the navigation was opened in 1831. Trade for Norwich poured in through the lock-gates of Lowestoft, and her old rival, Yarmouth was greatly disconcerted. But the unexpected happened. The sea which ushered in the traffic ushered in silt at the same time. In a few years the whole concern was ruined and bankrupt. A relic of it is the inner harbour. The outer harbour was built some time later, and this is the picturesque feature of Lowestoft. Here you may still see fishing smacks of the old order which have not adopted motors, great ketch-rigged vessels with brown sails and topmasts as tall as ever. There is far more business done both in the fishing and the herring-curing at Yarmouth, but the pageant of the trade is at Lowestoft.

But perhaps the most unexpected thing Lowestoft ever did was in the matter of receiving a sovereign of England. George the Second, on his return from Hanover, made for this port. The ship containing the royal entourage anchored in the roads. The King and his party, which included the Countess of Yarmouth, put off for the shore in a royal barge. Before they touched ground they were met by "a body of sailors uniformly dressed" who dashed into the sea and hoisted the barge and all its contents on to their shoulders and carried it up the beach, a feat which must have demanded enormous strength in the supporters and not a little nerve in the supported. But that was not the end of the adventure. The King was met by a Mr. Jex with his carriage. And Mr. Jex had no mind to let the rest of the reception be a tame one. He "conducted the King to his house, acting as coachman, and by his unskilful driving, narrowly occasioned the loss of his sovereign's life."

Between Lowestoft and Yarmouth the coast runs due north, clean cut, under a line of yellow-grey cliffs, showing no inclination to form bays or promontories. Yarmouth has no outer harbour. Its busy port is situated within the mouth of the River Yare. There is much shipping here besides that of the herring industry. But the importance of the herring to the town is still paramount. Yarmouth's origin probably goes back beyond the Saxon Period. It was then little more than a sandbank island at the mouth of the Yare. Here was held an annual festival called the Yarmouth Fair which was a forgathering of people and ships for the purpose first of fishing the herring and then of putting him in pickle for consumption during the winter months. At that date the only people who owned ships in a large way were the barons of the Cinque Ports. They brought their ships to the Yarmouth Fair, presided over it and instituted a collection of dues and tolls. In course of time a permanent settlement grew up there. But the barons kept all matters of law and administration strictly in their own hands, arriving from Kent at the time of the fair and presiding in state until the Cinque Port fleet was ready to return. The hall where they held their court is still to be seen.

But the place was no longer a fishing station. It was a growing town, while the Cinque Ports were gradually falling into decay. The rights of the barons became less and less as the Yarmouth people succeeded in fortifying themselves with royal charters. But they did not shake themselves entirely free of their old masters until the seventeenth century. In some ways the history of Yarmouth recalls that of Galway. Each was a microcosm of civilisation thriving in a remote place on the amenities of the sea. Each had its great families which were like tribes, each had an intimate contact with a foreign country—Holland on the one hand, Spain on the other.

Yarmouth has two immense waterfronts that are equally remarkable. At the back of the town is the long continuous quayside which is entirely dedicated to business. At the front of it is the sea esplanade (three miles long) which is entirely given up to pleasure. In this part, what with piers, pavilions, amusement parks, and the recently established sunk rock gardens, you might imagine yourself to be at any ordinary seaside resort. But there is still one happy reminder that you are not. That is the presence of the famous roadstead, a deep channel which passes along the whole of the marine front where shipping is protected from the violence of the high seas by the great shoals which lie parallel with the land. There is always a procession or a halting of steamers close to the water's edge. As many as one thousand five hundred ships have been seen at anchor at the same time in Yarmouth Roads. This gives the real romance of the sea to the holidaymaker while he takes his ease on the beach.

Since the failure of the original scheme to put Norwich in touch with the sea, the way has been cleared to reach this city both from Lowestoft and also Yarmouth. But it is only available to craft of very limited dimensions. These two ports are the gates of the Norfolk Broads, whose ramifications in river and mere take much longer to explore than a cursory glimpse at the map indicates. The glamour of the Broads is not easily communicable by verbal description. It depends equally on the beauty of shifting pastoral distances and on near foregrounds of reed, iris and water lily, all beheld under the spell of the gliding motion. The motor craft is a horrid intrusion into this erstwhile watery paradise.

The coastal margin of the Norfolk Broads is low. It extends roughly from Yarmouth to Happisburgh (pronounced Hazeborough). Intricate passages through outlying sandbanks lead into Yarmouth Roads and the perennial difficulties which

106 SALTHOUSE: a small Village on the North Norfolk Salt Marshes

107 KING'S LYNN, NORFOLK: the Restoration Custom House, by the local Architect, Henry Bell

these caused to masters of sailing ships produced a highly specialised sailor called the Yarmouth beach-man. Like the Deal hovellers the beach-men were pilots combined with salvage-men, but in this case salvage was the principal business. They worked by companies, had their look-outs and their huts where they kept a sharp eye on all shipping, day and night. Like the hovellers, they were an exceptionally bold and hardy race, putting to sea in all weathers to render aid to distressed vessels, performing amazing feats of boatmanship, and taking risks which no ordinary sailor would dare. It was the advent of steam tugs which brought about their dissolution. They had a tame but glorious ending. The last of these rough, fearless men took employment on the pleasure-beach of Yarmouth, among the bathing-machines, where they earned for themselves a unique reputation for courtesy to women and kindness to children.

The coast turns the shoulder of Norfolk at Cromer giving, at the same time, one of the most sudden surprises in scenery to be found anywhere along the margin of our five seas. It is caused by a ridge of hilly country which comes down to the coast and spreads itself out between Overstrand and Weybourne. Where the coast road rises by Mundesley you may look down over the Broads country and count perhaps a score of church towers rising among the trees all over that flat district. The level of the land and the level of the sea are divided as by a ruler along one sharp edge where the cliff runs in an unbroken straight line. Here, standing boldly out of the distance, rises the most glorious of all these towers, that of Happisburgh church, 150 feet high. We are in the threatened lands again, and it is significant that Happisburgh church has, as a companion tower, a cliff-top lighthouse. A large and important town must have stood where Happisburgh now is. The names of offlying sandbanks recall its further extension to the east.

It is not the height of the hills which lie behind Cromer and Sheringham which make that striking difference in the landscape between themselves and the fenland, for their highest point does not reach 300 feet. It is their shapes and the way things grow on them. The thick woods, the park lands, the heaths, the rounded bastions of bracken unite in giving a dynamic vivacity to the landscape which makes it react on the senses like Japanese scenery. I would not liken it to Japan except in the spiritual sense though

I think you could place the pagoda and the torii here without violating any feeling of fitness. It has not a true counterpart in all the range of British scenery.

The yellow front of Cromer Head descends from its 274 feet with a series of notches, the relics of landslips with which more than one lighthouse has gone down to the beach. From here the old town of Cromer with its red roofs and its magnificent church are seen to rise with precipitous grandeur from the waves against which it is embattled with strong walls.

Sheringham is the most developed and the most popular seaside resort along this part of the Norfolk coast. It is the home of a famous race of fishermen. The independent philosophy and the racy talk of these men couched in a dialect and an idiom, more pleasant to hear than any other, must have contributed largely to the success of the place.

The bird life of Norfolk has always been of outstanding interest. Such rarities as the hoopoe and the waxwing have been seen in the Sheringham woods. And nearly all members of of the great families of north-east Norfolk, while they led the commercial world in brewing and banking, were born naturalists. Between Sheringham and the Wash the coastline is broken by large salt marshes which, with their channels, project seaward behind long arms of shingle ridge. These are locally called harbours, the principal ones being at Blakeney and Brancaster. In both of them the National Trust owns land and foreshore rights where bird-watching stations and bird-sanctuaries are established.

The churches of Blakeney and Cley are among the finest in the country. Blakeney has, besides its fine western tower, a slender eastern one crowned by a lantern, made for the purpose of showing a light to mariners at sea. Cley church belongs to the height of the Decorated Period and the figures carved in the spandrels of its arches are among the triumphs of the sculpture of that date.

The waterfront of Wells is one on which you might stage an eighteenth-century pageant and find no discrepancy in its natural background. Its long channel winds out to sea through salt marshes and sandhills. Everywhere there are grand churches that must have been the centres of busy communities when wool was England's glory. Some are surrounded by pretty but humble villages, some are left alone in an empty countryside. Brancaster is a mere cluster of houses. But here a whelk fishery flourishes whose products go direct to the London stalls.

The coast turns in to the Wash at Hunstanton perhaps not so remarkable as a seaside resort as for the possession of variegated cliffs which are banded horizontally with red and white stripes. The coastline of the Wash is uneventful but not without beauty. The Sandringham woods fringe its eastern shores, and the fruit orchards of Cambridgeshire and Lincolnshire its southern margin. Its submerged sandbanks yield rich harvests of shrimps and small prawns to the fleets of trawling smacks which sail out of the Wash ports, and the banks which dry out afford a basking ground for seals of which there are great numbers in this small sea. The ancient town of King's Lynn with its still thriving port is in the south-west corner, at the mouth of the Ouse. Wisbech is reached up the Nene which goes to the sea in the middle of the Wash. Not very long ago the River Nene was navigable as far inland as the town of Northampton.

Wisbech is in the Cambridgeshire Fens, the country of Hereward the Wake. It was formerly a port on the coast of the Isle of Ely, but is now twelve miles inland. The transformation from fen into dry land took place in the seventeenth century under the guidance of the Dutch engineer, Vermuyden. The reclaimed land still goes by the two new names it acquired at that date, namely the Bedford Level and Holland.

The mouth of the River Welland is at the south-west corner of the Wash and gives access to Spalding. On the north-west coast, a little above the Welland, is the mouth of the Witham where shipping enters for Boston and Lincoln. This city is now only accessible by canal barge, but Boston's harbour, so important in the Middle Ages, and so hopelessly silted up in the eighteenth century, has been rejuvenated. Large steamers can now approach it and lie there in a spacious dock. Hull, Yarmouth, Boston, and Redcliff (Bristol), each claims, on various grounds, to have the finest parish church in England (abbeys excepted). If I were made judge, I would not hesitate to award the palm to Boston. And my award would not be based on the merits of its famous tower (the Stump) but on the effect of its interior on a stranger entering its western bay for the first time. Boston Stump is 263 feet high and is visible on a clear day for a distance of twenty-five miles. The old town with its ancient buildings and its winding water-front lined with shrimping-smacks preserves a dignity from its former greatness that in spirit is like that of Bruges.

The shores of the Wash are not spectacular, nor have they the same quality of old loneliness that I made note of in the Thanet meads and the mud-flats of the Thames Estuary. But this coast has a flavour of its own that strikes the imagination powerfully in storm, in fog and in the first weeks of spring when the sound of its numerous channels travels inland towards the great flats of the Bedford Level, where endless orchards load the air with the smell of blossom. The Wash has big tides, and the current of the tidal stream forms one immense eddy. The names of its sandbanks stimulate the bump of curiosity. Among them are Thief, Blackguard, Styleman's Middle, Roaring Middle, Westmark Knock, Roger, Peter Black, Pandora. And still the famous riddle as to where King John lost his treasure intrigues the world. As late as 1934 it was demonstrated in our courts of law how this riddle had so affected an American citizen that he had made its solution the object of his life. To this end he sunk many thousands of pounds in experts, typewriters, cameras, maps and motor-cars. All these proving useless, a psychic gentleman offered a much cheaper solution by divination, at six hundred pounds. But the idealist's financial nerve had failed. So our riddle remains to diffuse its atmosphere with a heightened prestige over the mysterious shores of the Wash.

Lincolnshire is full of interest for the antiquary. Its system of hills called the Wolds is little known to the general tourist. But its coastline is uncompromisingly flat, composed entirely of sandhills. It has two seaside resorts, Skegness and Maplethorp. Skegness is situated just on the seaward side of the mouth of the Wash. To it must be awarded the credit of an achievement which is perhaps unique in either landward or seaward resorts. It is capable not only of absorbing the tripper but of making him picturesque. You have, first of all, a town that is so flat that it does not overawe the pleasurer on the beach. Added to that, the beach is so roomy that there can never be any congestion. On this framework, which is a present from Nature, Skegness has improved with the greatest common sense. The spectacle on a hot day of high summer is one which any true artist could hardly fail to be thrilled with. Long beach-screens of red and white cloth stretch athwart the breeze to protect the non-bather from blown sand. Depressions and moist gullies are bridged by portable

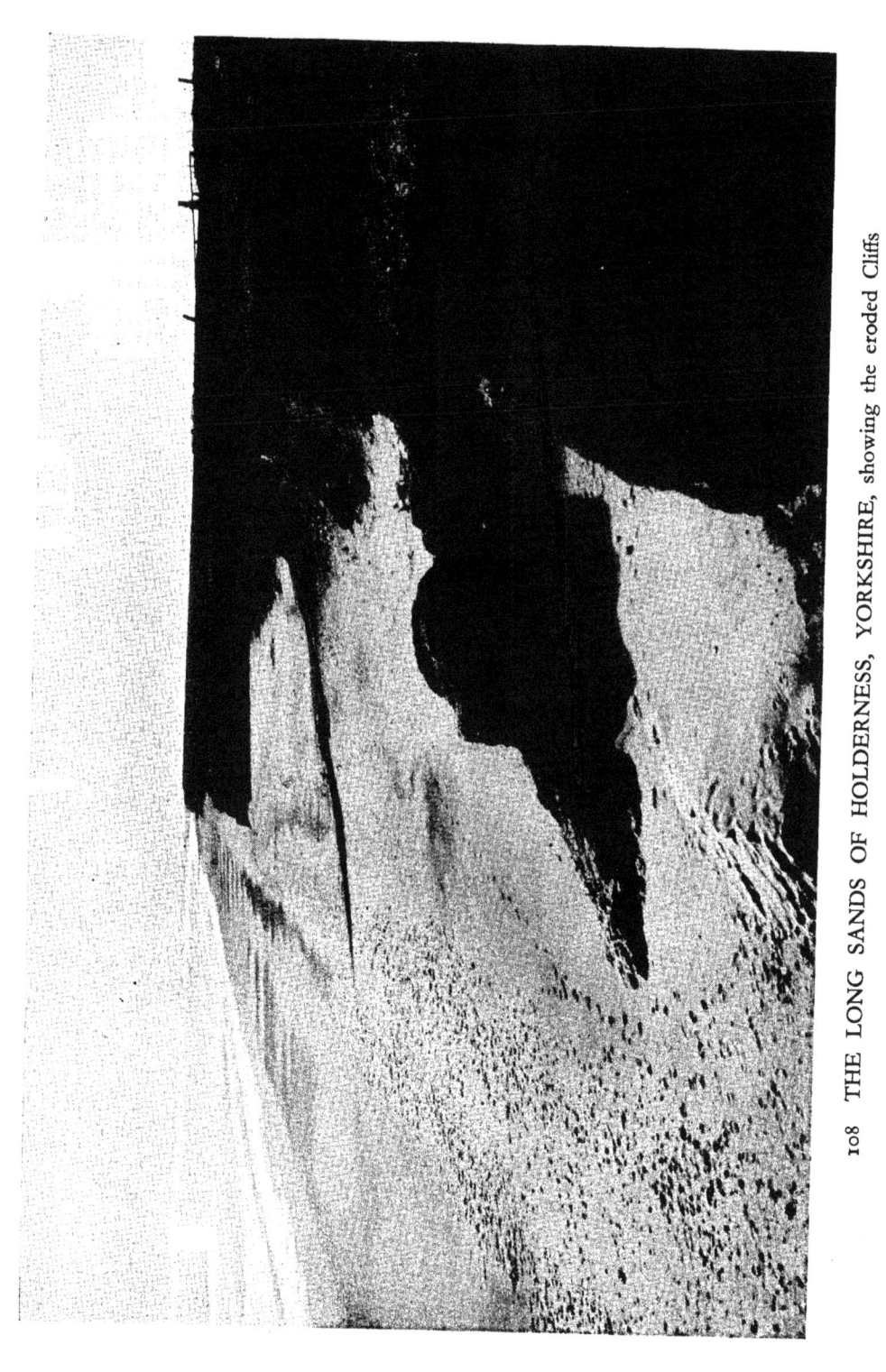

108 THE LONG SANDS OF HOLDERNESS, YORKSHIRE, showing the eroded Cliffs

109 SILEX BAY AND THE CHALK CLIFFS OF FLAMBOROUGH HEAD, YORKSHIRE

drawbridges made of duck-boards mounted on old motor wheels. Bathing vans with their horses follow the tide. At the edge of the sea a company of *bona fide* fishermen in blue jerseys and sea boots are strenuously engaged in managing the sea-going traffic. This is conveyed to the offing by a fleet of rowing-boats, motor boats, and steamers of a super-nautical-pantomime cut. The link between the shore and the main is provided by wheeled landing-stages and gaily coloured farm carts, from whose ends the wain-boards have been removed and large fenders bolstered on. These carts fill up with rejoicing family parties and are then hauled out to the shipping by splendid heavy-draft horses. The scene is further enlivened by stall-holders, Punch-and-Judy men, kite-fliers, and human shapes of all kinds in all colours of beach wear.

At the mouth of the Humber we are confronted by two curious geographical contradictions, for the Humber is not a river, neither is Hull a place. The Humber is really an estuary where the Trent, the Yorkshire Ouse, and the River Hull meet in confluence. In the thirteenth century, Edward I acquired the town of Wyke-upon-Hull from the monks of Meaux, fitted it out as a royal harbour, gave it a charter, and renamed it Kingston-upon-Hull. In the ordnance maps it still goes by this name. Elsewhere it is known simply as Hull. Its present greatness is due to the industrial development of Yorkshire in the nineteenth century. But up to that time it had had a long and important history. Like London, it had a rich sailors' guild called Trinity House. This has only ceased from its active maritime functions and become a charitable institution within recent years, its powers of pilotage, buoyage and lights having been vested in the Humber Conservancy Board.

The trade of Hull has always been intimately connected with Scandinavia, the Baltic and the Netherlands. The tower of its magnificent church of Holy Trinity is built of bricks said to be imported from the Netherlands at a time when this building-material was unknown elsewhere in England. Its early commercial supremacy was gravely threatened by Ravenspur, near the mouth of the Humber, but this place is now only a name found in history books in connection with the landing of Bolingbroke before he was crowned as King Henry the Fourth. Coastal erosion (which is reckoned to have advanced a mile and a half since Norman times) has swept it from the map. I wonder how many historical students have searched there in vain for it!

The Trent is canalised and large barges towed in strings by steam tugs come down from Nottingham. Fair-sized steamers can get up as far as Gainsborough, which is also visited by a tidal bore that has a unique name on this river—the *aiger* (pronounced as a rhyme to Hagar). The L.M.S. port of Goole is situated on the lower reaches of the Ouse where the Dutch River (a memory of Vermuyden) branches off. Goole is also the terminus of the Aire and Calder Canal, up which large barges with steam towage are conducted to Wakefield and Leeds. The Ouse is navigated as far up as York, and Selby receives that most silent of our services the fleet of H.M. Ordnance.

In the lower Humber, on the south bank, is situated the coal and general port of Immingham. It was opened in 1912 as a relief port to Grimsby. Both these ports owe their existence to the enterprise of the North-Eastern Railway. The first steam trawler was a converted paddle-tug owned by a Hull captain. His venture was much scoffed at by his friends and the owners of sailing trawlers. But even he can hardly have dreamed of the vast fleet of steam-trawlermen which go out daily from Grimsby to sweep the Dogger Bank. That old laughing-stock of a paddle-tug may have boded well for the consumer, the fishmonger, and certain great vested interests, but it boded no good for the local fishermen or for the harassed spawning and feeding grounds in all our seas.

The country between Spurn Head and Bridlington still keeps its ancient name of Holderness. Isolated between the Yorkshire Wolds and the sea, it has a lore of its own. It is a low, flat land of willow-bordered streams. In the great wool days its pastures must have been productive of some of that immense wealth of which we have a reminder in Beverley Minster. Its seaside resorts are Whittlesea, Hornsey and Bridlington.

North of Bridlington the land rises above a line of high chalk cliffs which form the north arm of Bridlington Bay and culminate in Flamborough Head. Bridlington has a busy little harbour and is blessed with a fine sandy foreshore. The name Flamborough is said to have come from the beacon which has flamed on that wild and dangerous headland since earliest memories of our shipping. The tower of a lighthouse built perhaps two centuries before the present one stands still, but the headland has probably been lit by a beacon of some sort since the time of the Romans. Flamborough has seen many remarkable epics of seafaring including that amazing fight between Paul Jones, the American privateer, and H.M.S. *Serapis*, which hauled down her colours when both ships were in a sinking condition.

The farmers of Flamborough have, from time immemorial, had a productive side-line in egg-collecting during the breeding season of the sea birds. Like the fowlers of the Shetlands they descend the sheer white cliffs on a rope, aided only by a confederate with a crow-bar. There is a good local market for gulls' eggs, and blown guillemot's eggs are sold to the summer visitor in Bridlington for a penny apiece.

At Bridlington the tripper element predominates. Filey entertains the more exclusive guest. A notable feature here is a reef of calcareous grit called the Brig which extends seaward. Filey, as a fishing port, owes its presence to the shelter afforded by this projection. But in the sailing-ship days it was one of the most dreaded perils of the North Sea.

In this part of the coast we take farewell of the three great banks of friable rock which stretch obliquely across England, the chalk, the lias and the oolite. To these we owe (as I mentioned when we greeted them) our southern and midland "downs," our supplies of concrete, and our most notable ecclesiastical building-stones. In the order that we have circumnavigated the coast the chalk was met at Beer Regis in Devon. In the striking, rugged, cave-riddled headland of Flamborough we part from it. We fell in with the lias at the blue cliffs of Lyme Regis, and with the oolite at Portland Bill. We part with these on this short but dramatic sector of the coast, between Filey and Redcar. Here they are more jostled together than in the south and not so easy to distinguish apart except to the trained eye of the geologist. But, as a general statement, it may be said that it is the lias which makes the cliffs while its younger companion, the oolite, builds the moors which lie immediately inland of them.

The farewell region of the oolite and lias is quite one of the most generally beautiful in all Britain. Yet it has never figured as a unit of territory before the days of tourism. To the tourist, however, it must appear as a single entity of topography, for it blends in neighbourly harmony all the elements he prizes most. It contains magnificent coast scenery (including our highest perpendicular cliff), moor and upland, woodland and deerpark, splendid ruins of the feudal days and some of the most beautiful of our villages. The district is roughly triangular. You may trace it through from the low ground of Filey up the Valley of the Yorkshire Derwent, past Pickering, and over the low watershed into the Vale of York; thence north to the estuary of the Tees; and so, coastwise, back to Filey. Within these bounds lie the moors of the Hambledon Hill and the Cleveland District with a group of high uplands to the south which do not bear a collective name. For the use and the delight of man it yields iron, alum, jet and the once marketable ammonite fossil.

One cannot but regard Scarborough as the metropolis of seaside resorts for it has unique natural advantages combined with an independence in wealth drawn from its home county and augmented by that of an immense migrant population from the north, the middle, and the south of England. Its great headland crowned by ruins of mediaeval and Roman masonry, together with its two sandy bays, one on either side, give it a beauty hard to defeat. Its town and sea-front buildings, though not beautiful in themselves, create a solid impression of spa magnificence, while the harbour, where the true business of the sea is transacted in full view, adds a final superlative note of the reality of romance.

North of Scarborough the cliffs rise to heights between four and five hundred feet. Here Ravenscar is situated. Its small beach is reached by a headlong descent. But the matter of easy bathing facilities has been solved by the local authority by building a swimming-pool on the cliff-top. The topography in these parts is concise. Interim features between the cliff and the beach are the *becks* (streams) and *brews* (hills). The small inlets on the coast which vary from coves to clefts are called *wykes*. The deep, steep, luxuriant valleys scored by the becks are filled thickly with woods and fern and present a striking contrast to the bleakness of the upland level. They are called dales. The actual moorland begins a little way back from the coast and reaches a tableland of between thirteen and fourteen hundred feet.

Beyond Robin Hood's Bay, with its wide indent into the high ground, comes the entrance to the Yorkshire Esk Dale. This fine, picturesque cleft where Whitby is situated admits of no estuary, only a good-sized tidal harbour, lined by quays on either side. The older part of the town is on the south bank, and is crowned by the splendid ruin of an abbey church, and by the parish church which cuts a figure from seaward as bold as a castle. The interior of this latter building must be

110 WHITBY, YORKSHIRE, looking to the East Cliff and Abbey Ruin

111 RUNSWICK BAY, YORKSHIRE, flanked by its tall Cliffs

a little disappointing to archaeologists but contains much to charm the lover of the quaint and curious. Its woodwork has been devised by ships' carpenters, and on the side of its Georgian pulpit are fastened two ear-trumpets, instruments whereby the deaf but devoted wife of a late incumbent was enabled to hear the sermons of her spouse. Could wifely adoration go further?

Formerly, Whitby had a prominent industry in shipbuilding and furnished ships to Captain Cook for his voyages of discovery. When we remember that it was only through the stoutness of that vessel which struck the Great Barrier Reef of Australia that Cook survived to translate his finds into the facts of history, we may well thank the slipways of Whitby for our sovereignty in Australasia. The small but interesting industry of making ornaments out of the local jet has survived the withering of the Victorian widows' weeds. The port is lively with the activities of fishermen and visiting shipping, but the building stocks have long been idle. Fortunately the close-clustered, red-roofed old town is not disturbed by all the modern arrangements made for the summer visitor trade. These are developed along the cliff to the north. Here I have one thing against Whitby. How is it that the local authority has come to spend fabulous sums on such an extravagant amenity as a lift up its cliff face and yet not seen to it that the town had a first-class restaurant?

At Sandsend Wyke the cliffs are grey-blue and riddled with the caverns of old alum works. The alum quay, with its massive walls and huge worn iron rings, to which trim brigs and schooners were made fast during loading operations, survives to add an exclamation mark of glamour to the end of a small marine parade. The old men of the district still remember the exigencies of the alum trade. How the boughs of the neighbouring oak woods, mostly wind-felled, called *garzel*—which I fancy is an old forestry term—were brought down to the quay and carefully peeled of their bark. The oak bark went to the local tan yards (there was life in the leather of those days) and the wood was used to burn the alum shale. Large liasic concretions called *doggers* (after which the trawling bank is named) were also found here and burnt for Roman cement.

Runswick Bay has a splendid sweep of sand set between two magnificent headlands whose general russet colouring is heightened in places by red and orange tinges. When they are floodlit by sunset and a sleek swell throws up a margin of glistening rose-tinted ivory the spectacle is a rich one.

Staithes is built in a cleft, much narrower and more imposing than Whitby's, carved out by the joint action of several becks. It is probably the most genuinely "old-world" fishing village left on our coasts. I think the truth that its quaintness is not a cult wilfully persevered in to draw the summer visitor is established by the fact that its young women continue to wear the sun-bonnet, that delightfully picturesque headgear which, in general, has passed into the category of "costume" or is used only by the old wives who have more sense than vanity. Every bit of Staithes is picturesque, its shop-fronts, its river-harbour, its waterfront whereon stands the Cod and Lobster Inn, tarred all over, wall and roof, and equipped with special shutters to its windows, because the seas break right over it in the winter-time. This is not a place where among the fisherfolk "men may work and women may weep." Here the women spend the day baiting the long-lines which their husbands take out to sea with them at night. A long-line may be as much as three miles long, mounted with hooks only a yard apart from each other. I wonder whether the visitor who comes here in the pleasant days of summer ever pictures to himself what the place is like in winter, the season of long-lining, when this little town in a reverberating cleft is deafened with the bombardment of the seas and blinded with the flying spindrift!

That great sea precipice 672 feet high (which I think is the highest in England) called Boulby Cliff by the landsman, and Redcliff by the sailor, stands over the north entrance to Staithes. From half-way up its face a small cloud of steam like the smoke of a volcano issues perpetually, caused, they say, by a chemical reaction in the alum shale. The hollow of a neighbouring cliff is full of strange grottoes and of the fragments of ruinous buildings. These, you will hear, are the works of the Romans, by which you must understand the men who sought for and wrought "doggers" into Roman cement.

But works of another kind, having immense buildings and chimneys, with attached villages, smirched with the squalor that attends all operations in iron and coal, possess all the coast and the interior between Staithes and Saltburn. They are the outlets for the native iron of the Cleveland District. Saltburn is built on the cliff top and has an organised sea-front on the

112 THE CLUSTERED ROOFS OF STAITHES ON THE
YORKSHIRE COAST

113 BRIDLINGTON, YORKSHIRE: the Fishing Harbour

beach below. Its neighbour, Redcar, is wholly on the level. It is a resort of the Blackpool order.

The mouth of the Tees is the entrance to the most industrialised area of the East Coast's Black Country, where the combined products of the Yorkshire iron mines and the Durham coalfields are "made up" for use at home and abroad, or exported in the raw. The estuary becomes a river at Middlesbrough which, in 1820, had only one house (still shown), but by 1921 had acquired a population of 131,103 persons. Here ship-building and iron-and-steel are the ruling trades. Navigation is continued to Stockton-on-Tees, whose name coupled with that of Darlington, has first place in the history of all railways. At the northern corner of the Tees mouth, in a bay of its own, is Hartlepool. In the Middle Ages it was a fortified town and port belonging to the prince-bishops of Durham who are said to have owned galleys and kept them there. It still preserves the pronunciation of the Norman French *le* in the middle of its name. To-day it has immense docks and a large steam-trawler industry.

The Durham coast is composed of cliffs of grey magnesian limestone. But, from the tourist's point of view, it is a wholly black coast. Its ports of Seaham and Sunderland are given up to coal, as are the intermediate villages. Before the bishops of Durham parted with their rights and privileges of being princes of a county palatine, the wrecks along this rugged coast formed an appreciable part of their rich revenue, and shore bailiffs were maintained to look after the spoils of the sea. I am always inclined to believe that the tradition that the captain of a ship should not leave her, even at the risk of drowning, has a basis of expediency rather than heroism. For in the old days if a ship went ashore without a responsible person on board she became the property of the lord of the manor. But if the master was still on board she could not be claimed, and might be salved.

The boat-building on this coast, especially north of the Wash, is quite distinctive from that on the west and south coasts. The most peculiar and interesting of these types is the coble of Yorkshire and Northumberland—I do not think there is any appreciable difference between the two except that the *o* is pronounced short in Yorkshire and long in Northumberland. She is a keel-less boat, rounded in the bows and flat in the stern. The object of this is to enable you to make a safe stern-on landing on a sandy beach in rough weather.

For purposes of sailing, the lack of keel is made up for by a very deep rudder which projects far below the draught of the boat and has therefore to be quickly unshipped at the moment of landing. Sailing the coble efficiently was a great and noble art, but, since the coming of motor-engines, hardly one sailing coble is to be found on the whole coast. The young generation will not learn, and it is not hard to predict a speedy disappearance of these picturesque boats from the beaches. For the elaborate hull-construction of the coble is unnecessary when sail is not carried; economic adaptations for the motor will set in as soon as the boat-builder can get the better of the conservativeness of the fisherman.

The Tyne has an imposing entrance. To the south is a grand sweep of rich yellow sand, to the north a steep promontory of yellow rock where stand the ruins of a large castle and a priory. There, too, looms a battery of more modern date, while, behind, from among the quiet verandahed houses of the Victorian spa, rise the column carrying the figure of Admiral Collingwood and one of the older and quainter types of lighthouses. Two curved piers, one a mile in length, embrace the river as it goes out to sea for the purpose of focusing the scour of its freshets on the bar.

The river banks (especially the north one) are scenes of activity nearly all the way to Newcastle. The north bank is occupied with shipyards and docks and the south with coal staithes. *Staithe* is a word which belongs to the seaboard between the Tweed and the Humber. It means, in general, a landing-place, and, in particular, one built on high wooden legs. On the Tyne, the coal-pits are close to the water's edge, and the pit-heads are situated on rising ground which enables trains of trucks to be lowered on wire ropes from the collieries to the staithes, and to be pulled up again, without the use of locomotives. On the staithes, which are raised high above the decks of the craft alongside, the bottoms of the trucks are knocked open, and the coal hurtles down shoots into the ships' holds.

The Tyne is a confluence of two rivers of the same name—North and South Tyne—which unite thirty-six miles from the sea. They come from country that is intensely romantic both in scenery and history, flowing by Hexham and the Roman wall of Hadrian. Even commercial Newcastle still salutes the Tyne with grand and ancient buildings—a splendid Norman keep and the tower of St. Nicholas Cathedral, which is a link between Boston Stump and the tower of St. Giles at Edinburgh. Even amidst the unutterable grime and squalor of Jarrow, the Saxon remains (and they are substantial ones) of the monastary where the Venerable Bede penned his history stand in a little quiet oasis at the edge of the tidal muds.

The shipbuilding industry of Tyneside is of modern origin, dating from the mid-nineteenth century, when the piers at the mouth were built and the river deepened. Previously, the estuary was a very shallow one and the trade was confined to the export of coal. Coal from Newcastle (and also from South Wales) has been taken by ships to ports all round the coast since Norman times, for which reason this fuel is always mentioned in old documents as *sea-cole*. The trade was suspended temporarily under a ban of Edward I who feared for the safety of old wooden London. The Newcastle collier was called a *keel*, a craft which has only passed out of existence since the War. The keel-men were a powerful party and formed a gild of no mean order. To their memory and the Tyneside dialect we owe the famous snatch—

"Weel may the keel row."

The modern Tyneside sailor is called a Geordie, but I have not found out why.

The Northumberland coast has a distinct flavour of its own. Compared with the richness, both wild and luxuriant, of its inland scenery, it is uneventful. But wherever you go along that low shore there is always something exceptional to give the setting a quality that can only be called Northumbrian. It may be sweet briar among the sandhills, or cobles drawn up in a particular way by a fishing village, or castle built on basalt rock, or a glimpse at the blue outline of the Cheviots, or just a certain sparkle of the clear sea water which rolls in unobstructed by outlying shoals, or a taste in the air, or a gem of local cloudscape. Whatever it is, it is individual and romantic, if slightly melancholy (for the Northumbrians have not that forceful optimism of the Yorkshiremen). But the charm does not begin to work till the coalfield ends.

In the meantime, the yellow sandstone rock of the coal measures provides a richly-coloured and altogether delightful foreshore all along the coast. Tynemouth is mellowed by memories of its Victorian prime. Whitley Bay is pervaded by the architecture and sentiments of the rising generation. Cullercoats preserves its fisherfolk untouched. Here you will see women sitting in their doorways selling dressed crabs, freshly cooked lobsters, and shrimps by the pint. Would that more of our fishing villages did likewise! To go to a seaside resort and not to be able to eat the fresh yield of the sea is as bad as to go into the country and have to eat crate eggs. Yet how rare it is to find such facilities as at Cullercoats! Blythe is a port of coal-staithes having a long detached promenade with appropriate buildings among its adjacent sand-hills, erected for the entertainment of the holiday-making collier. Newbiggin is a similar compound of colliery and sea front.

The coalfield ends finally at Amble. And where it ends, the romantic note sets in with a flourish. Amble Harbour is on the mouth of the River Coquet. Coquet Island with its white lighthouse and slight remains of a religious settlement of the very early days of Christianity lies just off the entrance. At Amble there is a boat-builders' yard, I think the only one left in Northumberland which builds cobles. The river flows down to meet the tides through a winding defile thickly embowered with trees. It has a bed of ledges, over which the water alternately waits and frets. Between tree-root and water the margins are succulently and slumbrously garnished with the plate-like leaves of the butter-burr. At a bend towards the first deep ravine the old humped bridge of Warkworth crosses it. This bridge keeps its mediaeval gatehouse. If you pass through it you will see the wide street going straight uphill through the village to the great castle on top where the immense figure of the lion rampant of the Percys is displayed on the wall of the keep. From a high, slender watchtower above that massive building you overlook the port of Amble, and Coquet Island, and the sea horizon far beyond.

It is a far cry round the coast to the great castles of Wales, but it is nearer that way than overland in times of trouble. Messengers to and from the Welsh shores must have slipped in and out of Amble in 1403. For in that year the two castles of Harlech and Warkworth were knit together in a strange bond. Owen Glendower who had taken Harlech had given his daughter in marriage to Roger Mortimer, and Mortimer's young aunt was the wife of Harry Hotspur. It was from Warkworth that Hotspur set off to meet the King at Shrewsbury. Within how short a time of his departure was he to stand naked and dead between two millstones at the pillory in Shrewsbury market!

114 SHIPPING IN THE TYNE AT SOUTH SHIELDS, COUNTY DURHAM

115 THE RUIN OF TYNEMOUTH PRIORY, NORTHUMBERLAND

116 BAMBOROUGH CASTLE: a famous Stronghold of the Northumberland Seaboard

117 DUNSTANBURGH CASTLE, NORTHUMBERLAND, from the South

After the Coquet comes the Alne. It makes a pretty loop in low-lying ground before rounding a barrier of rock to go to sea. On this barrier stands the town of Alnmouth, having the river on one side and the sea on the other. As a resort it is remarkable for having its "front" situated on neither river nor sea but, informally, on the main street between the two. The salmon fishers here have not only to pay the usual exorbitant licence to the Government but an additional fee of a pound to the lord of the manor for use of the foreshore.

Boulmer and Crastor are two fishing villages which are more given up to their legitimate business than attempting to "cater for" visitors. Crastor has curing houses ready for the herring when the shoal arrives. The old-fashioned method of smoking with oak sawdust *only* is used, and the Crastor kipper has a wide reputation locally. Crastor's natural harbour is bitten out of a lump of black columnar basalt. It is not so regular in its honeycomb cubes as at Staffa and the Giant's Causeway. None the less, it makes natural colonnades sufficiently striking.

This handsome adamantine volcanic rock which first appears on the coast at Crastor makes numerous and always noteworthy interjections between here and the Tweed. It is the end of a long narrow belt called the Whin Sill which first crops out at Haltwhistle on the edge of Cumberland and shows at intervals in an oblique line towards the coast. At its first appearance where it makes a steep and imposing precipice facing north, it is crowned by the ruin of Hadrian's Roman wall. Indeed that great work derives half its grandeur, and the hold it has on the imagination, from its partnership with the Whin Sill. We are now to observe three other striking combinations of this rock with the works of man.

A little beyond Crastor the Whin Sill bulges from the sea's bed just at the land's edge. On this promontory stands Dunstanburgh Castle, an immense ruin whose light yellow stone contrasts tellingly with the black of the basalt and the sparkling blue clearness of the sea. When a gleam of sun catches it against a piled-up background of cloud it makes a picture more startling than even Turner could have achieved.

And there are things in Dunstanburgh's story that are as romantic as its appearance, for it is associated with three of the most picturesque rebels in English history. Its site was owned by Simon de Montfort—another curious link with the romantic coast of Wales, for his daughter married Llewelyn, the last native prince. But the castle was built a little later by Thomas of Lancaster. Like de Montford, Thomas was a great rebel though not a great idealist. He was the only person who ever drew a manful stroke from Edward the Second. He is said to have laughed at the king from the security of his battlements at Pontefract when the beaten army from Bannockburn passed by, and he connived at the capture and murder of Gaveston, the royal favourite. He had not time to finish the building of Dunstanburgh before Edward met him at Pontefract and had him beheaded in his own dining-hall. Guy, Earl of Warwick, was the third notable. He did Dunstanburgh the honour of feasting in it after he had effected its capture. The sailors have a descriptive name for its wild, jagged ruin as it appears from the offing. They call it the Snags of Dunstanburgh.

Sea Houses is a busy fishing-port with a very picturesque harbour and water-front. Out to sea, fragments of the Whin Sill rise up and form the Farne Islands. Here, the Longstone lighthouse was the home of Grace Darling and her father who rowed out in their coble in the teeth of a fierce gale to save the crew of a wrecked steamer. The Inner Farne has a longer memory, recalling the life of St. Cuthbert. These islands are now owned by the National Trust as a bird sanctuary. The eider duck breeds here.

A few miles to the north of Sea Houses the Whin Sill raises a flat-topped precipitous hill, this time just within the shore-line. It is crowned with the *chef d'œuvre* of all our castles. I do not know any building in all the British Isles so unbelievable to behold at first sight as Bamborough Castle. Its immense curtain walls rising from the cliff face in which are set towers round and square, its inner buildings with their roofs and turrets leading up to the crowning majesty of its Norman keep strikes the imagination a real home thrust. Its history is a lively one. Many famous men and women, kings, queens, nobles and prelates have figured in its story. It has stood a number of sieges, including a famous one by Warwick the King-maker. Restorations began in the eighteenth century and have continued up to the present day, when the building has become so modernised as to be let off in flats and "quarters." So there is much that the antiquarian has to bemoan, and the near view causes some disillusionment. But as an impression, as an accent in a great coastal landscape, Bamborough Castle has not its match.

The proper setting of Bamborough bears away to the north, not to the south, as we have approached it. Here lies Lindisfarne or Holy Island at the entrance to a deep bay whose vast area of sand dries out at low tide. The island has its ruined abbey, its ancient church and its castle, perched on the last and, in some ways, most striking freaks of the Whin Sill. The main approach to it is from Beal, at the low tide. The distance over the sands is three miles, and two routes are marked out with tall stakes which have at intervals safety-boxes mounted on them, into which a traveller who is caught by the sea can clamber for safety. Ancient motor-cars are used to convey the summer visitor over the sands. But this, to my mind, is not in accordance with the charms of the island and its association with holy calm and detachment. A better way is to make your pilgrimage along the trackless warren which juts out as a peninsula to the south side of the bay, and approaches Lindisfarne to within a couple of hundred yards. Here you may make a signal and one of the ever-watchful islanders is sure to see you and bring his boat over. But even here you must have an eye to the state of the tide, for the end of the peninsula is cut off for about two hours on either side of high water.

Notices are posted on the roads leading into Berwick-upon-Tweed informing you that it is Britain's most historic town and advising you to linger there and inspect its ancient buildings. But what historic buildings are there to see there except the Elizabethan town walls built by an Italian architect and a few fragments of the Edwardian ones? There is certainly an attractive atmosphere of ancientness in the town. But that ancientness is not tangible (with the exceptions mentioned) below the eighteenth century. The local authority has done away with nearly everything to remind us of Berwick's greatness in the past. And in spite of its notices its attitude towards old things is still unregenerate. Even its museum contains hardly any local relic of importance.

But if you climb to a point on the high ground on the south bank of the Tweed the town with its red roofs, its towers and spires, its crooked hillside streets, its bridges (where you may count forty-three arches), its embattled girdle of white stone, and its harbour wall which has its outer footing in the open sea—if you stand watching this picture carefully, you may easily restore the vision of that town which had a separate entity between England and Scotland. And if you turn to look back you will see all the way down to Lindisfarne, and past it to Bamborough in its majesty, while the sea horizon is faintly punctuated with the tower of Grace Darling's home, the Longstone lighthouse.

INDEX

The Numerals in italic type denote the *figure numbers* of Illustrations

Abbotsbury, 75
Aberayron, 43; *40*
Aberdaron, 41
Aberffraw, 39
Aberporth, 44
Aberystwyth, 43
Alde, 92; *93*
Aldeburgh, 92, 93
Alnmouth, 113
Amble, 112
Amlwch, 34
Andreas, Plain of, 13, 16
Anglesey, 5, 23, 26, 28, 32–34, 39; *31*
Appledore, 63; *49*
Ardglas, 11
Arnside Knot, 20
Athelney, 62
Avalon, 62
Avon, River, 60, 61

Babbacombe, *72*
Balbriggan, 12
Bamborough Castle, 114, 116; *116*
Bangor, 31, 32
Bardsey, 41, 43
Barking, 93
Barmouth, 42; *36, 37*
Barnstaple, 63
Barrow-in-Furness, 18; *19*
Barry, 59
Beachy Head, 80; *89*
Beaumaris, 26, 28, 32
Bede, the Venerable, 94, 111
Bedford Level, the, 101, 102
Beer, 73, 105
Beesands, 68
Bellan Fort, 40
Benllech, 33
Berwick-upon-Tweed, 115
Bideford, 63
Birkenhead, 25
Bishop's Ooze, 88
Black Coombe, 18
Blackpool, 22; *20*
Blakeney, 100; *101*
Blythe, 112
Boat-building, 109, 112
Bolton-le-Sands, 20
Borth, 43; *34*
Boscastle, 66

Boston, 101, 111
Boulby Cliff, 108
Bournemouth, 77; *80*
Brancaster, 100
Bridgwater Bay, 62, 63
Bridlington, 104, 105; *113*
Bridport, *78*
Brightlingsea, 90
Brighton, 79; *86*
Bristol, 24, 50, 61, 77; *47*
Brixham, 72; *69, 73*
Broadstairs, *85*
Bude, 66; *63*
Bull Bay, 33
Bull Point, *48*

Caldy, 53
Calf of Man, 13
Camborne, *61*
Cardiff, 59, 60
Cardigan, 45; *39*
Cardigan Bay, 38, 41
Carleon, 61
Carlisle, 16, 17
Carmarthen Bay, 52–54, 56, 57
Carmel Head, 34
Carnarvon, 31, 38, 39, 40; *30*
Castletown, *16*
Chanctonbury Ring, 79
Charmouth, 74
Chesil Beach, 74
Chester, 24, 25, 26, 27, 36
Chichester, 78
Chicken, the, 13
Chops (of the Channel), 71
Christchurch, 77
Cinque Ports, 24, 80–82, 97
Clacton, 90, 91
Clarach Bay, *38*
Cley, 100
Cliftonville, 90
Clovelly, 64; *51*
Clwyd, 28
Colchester, 87, 90, 91
Colwyn Bay, 29
Connah's Quay, 26
Conway, 29, 32
Coquet Island, 112
Craigneish, 13
Cranforth, 19
Crastor, 113

117

Criccieth, 40
Cromer, 99, 100; *105*
Cullercoats, 111

Dartmoor, 68, 72
Dartmouth, 68, 72; *70*
Deal, 7, 80, 82, 83
Dean, Forest of, 18
Dedham, 92
Dee, River, 10, 15, 24–27, 62; *22*
Dogger Bank, 72, 104
Douglas, 13, 14
Dover, 72, 80, 81, 86; *88*
Dovey, River, 43
Duddon, River, 17, 18
Dulas Bay, 33
Dundrum, 11
Dungeness, 72, 81
Dunstanburgh Castle, 113; *117*
Dunster, 63
Dunwich, 94, 96; *98*
Dutch Settlements, 89, 91

Eastbourne, 80
Eddystone Rock, 50, 51, 68, 70
Elenith Hills, 56
Essex, 90

Falmouth, 68
Farne Islands, 114
Felixstowe, 92
Filey, 105
Fishguard, 45
Flamborough, 104, 105; *109*
Flatholm, 60
Fleetwood, 21, 22, 52
Flint Castle, *23*
Folkestone, 72, 80, 81
Foreland Point, 63
Foryd, 28
Fossils, 74
Fowey, 68
Freshwater Bay, *52*
Furness Abbey, 18
Fylde, The, 22, 23

Giltar Point, 52, 53
Ginst Point, 54
Glamorgan, Vale of, 58
Goodwick, 45, 46
Goodwins, The, 82, 83
Goole, 104
Gower, 54, 57
Gravesend, *92*
Great Orme, the, 28, 29, 33
Greenwich, 89
Grimsby, 104
Gynfelin Patches, 42

Hadrian's Wall, 113
Hakluyt's *Voyages of English Seamen*, 91
Halton Castle, 23
Hanseatic League, 84, 96
Happisburgh, 98, 99
Harlech, 42; *35*
Harlech Castle, 112
Hartland Point, 50, 64, 65
Hartlepool, 109
Harwich, 91, 92
Hastings, 80; *5, 91*
Helford, 68
Heysham, 20, 21; *14*
Hilbere, 26, 28
Holderness, 104; *108*
Hollesley Bay, 92
Holy Island, *see* Lindisfarne
Holyhead, 5, 26, 35, 36; *12*
Hook's Nose, 72
Hoylake, 25
Hull, 101, 103
Humber, the, 103, 104
Hunstanton, 101
Huntsman's Leap, 52
Hythe, 80

Ilfracombe, 63; *52*
Immingham, 104
Ipswich, *104*
Isle of Man, 12–15; *15*
Isle of Wight, 77, 78; *81*

Kidwelly, 57
Killoran, the Barque, 7
Kingsley, Charles, 26, 63
King's Lynn, 101; *107*
Kingstown, 12; *13*
Kircudbright, 15

Ladram Bay, *66*
Lancaster, 20
Land's End, 65; *55*
Langstone, 78
Laugharne, 55, 56
Lincoln, 101
Lindisfarne, 115, 116
Little Eye, 26
Liverpool, 24; *21*
Liverpool Bay, 10, 23
Lizards, The, 5, 67
Llanddwyn, 39
Llandrillo-yn-Rhos, 29
Llandudno, 28, 29; *25, 26*
Llanelly, 57
Llanfairfechan, 30, 43
Llangranog, 2, 44
Llanstephan Castle, 55
Lleyn, Peninsula of, 40, 41

INDEX

London, Port of, 9, 24, 87-89
Longstone lighthouse, 114, 116
Lowestoft, 67, 96, 97, 98; *9*
Lulworth, 76; *76*
Lundy Island, 59, 63
Lune, River, 20
Lydd, 80
Lyme Regis, 73, 74, 105; *74*
Lynmouth, 63; *2, 50*
Lynton, 63

Macaulay, *quoted*, 16
Maldon, *99*
Manorbier, 52
Maplin Sands, 90
Margate, 86
Marlowes, 48
Maryport, 16
Medway, the, 87, 88; *94, 95*
Megalithic monuments, 66
Menai Straits, 30, 31, 32, 39, 40; *29*
Meols, 25
Mersea Island, 90, 93
Mersey, the, 23, 24, 25, 27
Middlesbrough, 109
Milford Haven, 46, 47, 49, 51, 52
Milnethorp, 19
Moelfra, 33
Morecambe Bay, 10, 18-22; *18*
Mostyn, 26
Mourne Mountains, 12
Mumbles, The, 57; *45*

Neath, 18, 58
Needles, the, *81*
Nendrum, 11
Nevin, 40; *32*
New Brighton, 25
New Quay (Wales), 43, 44
Newbiggin, 112
Newborough, 39, 40
Newcastle, 110, 111
Newlyn, 67
Newport, 45, 61
Newquay (Cornwall), 66
Nith, 15
Norfolk Broads, the, 98-100
North Downs, the, 79, 81

Oddicombe Beach, *71*
Ore, 92
Orford, 92; *102*
Ormskirk, 23

Paignton, 72
Parkgate, 25
Pembroke Castle, 52
Penarth, 59, 60
Pendine, 54

Penmaenmawr, 30
Penzance, 67
Perranporth, *60*
Pevensey, 80
Plymouth, 68-71; *65*
Plynlimmon, 43, 44, 56, 61
Point Lynas, 5, 23, 27, 33
Point of Air, 26, 27, 28
Polperro, 68; *64*
Poole Harbour, 77, 78; *3*
Porchester, 78
Port Eynon, *46*
Port Isaac, 66
Port Penrhyn, 32
Port Talbot, 58
Porth Leven, *6*
Porthloe, *59*
Portland, 72, 75; *75*
Portmadoc, 41; *33*
Portmerion, 41
Portreath, *57*
Portsmouth, 78; *84*
Prescelly Mountain, 45, 47
Prestatyn, 28
Preston, 23
Puffin Island, 33; *27, 28*
Purbeck, 76, 77
Pwllheli, 40

Quantock Villages, 63

Ramsgate, 82
Ravenglas, 17
Ravenscar, 106
Ravenspur, 103
Reculver, 5, 86, 89; *97*
Redcar, 109
Redcliff, *see* Boulby Cliff
Redwharf, 33
Rennie, 26, 35, 36
Rhinog Range, 42, 44
Rhoscolyn, 39
Rhyl, 28; *24*
Ribble, River, 22, 23
Rochester, 88; *94, 95*
Rocksavage Abbey, 23
Romney Marsh, 81, 87
Rudyard, John, 70
Runswick Bay, 107; *111*
Ryde, 78
Rye, 80; *87*

St. Alban's Head, 76
St. Anne's Head, 49, 50, 51
St. Bride's Bay, 48
St. David's, 47, 48, 57
St. David's Head, 38, 43, 48
St. Donat's, 59
St. Govan's Head, 52; *44*

St. Ives, 66; *62*
St. Michael's Mount, 67; *56*
St. Osyth's, 96
Saltburn, 108
Saltcombe, 68; *67*
Salthouse, *106*
Sandwich, 80
Sarah Evans, Wreck of the, 10, *11*
Sarn Badrig, 42
Saunton Sands, *53*
Scarborough, 106
Scaw Fell, 17
Sea topography, 4–9
Seascale, 17
Selsey Bill, 79
Seven Sisters, the, 80
Severn, River, 43, 53, 60, 61, 62
Sharpness, 60
Sheringham, 99, 100
Shoeburyness, 90
Sidmouth, 73
Silloth, 16, 19; *17*
Skegness, 102
Skerry Islands, 34
Skiddaw, 16, 17
Skomer Island, 48
Smeaton, 20, 71, 82
Smuggling, 15, 59, 67, 93
Snowdon, 28, 41, 44, 79
Sole Bay, 95
Solent, the, 72, 77; *82*
Solway Firth, 15, 16, 19, 26, 53, 65, 80, 87
South Downs, 79, 80, 81
South Shields, *114*
Southampton, 24, 72, 77, 78; *83*
Southend, 90
Southport, 23
Southwold, 93, 95
Stackpole, 52
Staithes, 108; *112*
Start Point, 67, 71
teepholm, 60
Stockton-on-Tees, 109
Stour, River, 92
Strumble Head, 45, 47
Swale, River, 87
Swanage, 77; *77, 79*
Swansea, 18, 58

Taw, River, 63
Teignmouth, 72
Telford, 26, 31, 35, 61, 62, 96
Tenby, 53; *43*
Thames Estuary, 79, 86, 87, 92; *92, 93*
Tintagel, 66; *58*
Towyn, 43
Traeth Saeth, 44; *42*
Turner, 10, 25, 71, 113
Tyne, River, 110
Tynemouth, 111; *115*

Ulverston, 18
Usk, River, 60, 61

Valency, 66
Ventnor, 77, 78
Vermuyden, 89, 101, 104

Walmer, 80
Wantsume, 86, 87
Warbarrow Bay, 76
Warkworth Castle, 112
Warton, 20
Wash, the, 100–102, 112
Wells (Norfolk), 100
West Kirby, 25
Weymouth, 73, 75
Whin Sill, 113, 114
Whitby, 74, 106, 108; *110*
Whitehaven, 16
Whitesand, 67
Whitesides, H., 51
Whitley Bay, 111
Whitstable, 87
Widemouth Bay, *54*
Wilkinson, 18
Winchelsea, 80
Winstanley, 51, 70
Wirral, the, 25, 28
Wisbech, 101
Woodbridge, *100*
Workington, 16
Wren, Sir Christopher, 75
Wye, River, 43, 60, 61

Yarmouth, 67, 79, 80, 96–99, 101; *103*
Yorkshire Esk Dale, 106

Vale, Edmund, 1888–
    The seas & shores of England, by Edmund Vale; with a foreword by Sir Arthur Quiller-Couch ("Q") ... New York, C. Scribner's sons; London, B. T. Batsford, ltd., 1936.
        viii, 120 p. col. front., plates. 22 cm. [The English countryside series]

    "Printed in Great Britain."

1. England—Descr. & trav.—1901–1945.   I. Title.
                        Full name: Henry Edmund Theodoric **Vale.**

DA630.V28   1936a   914.2   36–27309

Library of Congress   [a62f½]